The International Politics of Genetically Modified Food

Also by Robert Falkner

The Cartagena Protocol on Biosafety : Reconciling Trade in Biotechnology with Environment and Development? (Co-edited with Christoph Bail and Helen Marquard)

The International Politics of Genetically Modified Food

Diplomacy, Trade and Law

Edited by

Robert Falkner

Lecturer in International Relations, London School of Economics

363.1929
I61

First published 2007 by
PALGRAVE MACMILLAN
Houndmills, Basingstoke, Hampshire RG21 6XS and
175 Fifth Avenue, New York, N.Y. 10010
Companies and representatives throughout the world

PALGRAVE MACMILLAN is the global academic imprint of the Palgrave
Macmillan division of St. Martin's Press, LLC and of Palgrave Macmillan Ltd.
Macmillan® is a registered trademark in the United States, United Kingdom
and other countries. Palgrave is a registered trademark in the European
Union and other countries.

ISBN-13: 978-0230-001251 hardback
ISBN-10: 0-0230-001254 hardback

This book is printed on paper suitable for recycling and made from fully
managed and sustained forest sources.

A catalogue record for this book is available from the British Library.

Library of Congress Cataloging-in-Publication Data

The international politics of genetically modified food : diplomacy, trade, and
    law / edited by Robert Falkner.
        p. cm.
    Includes bibliographical references and index.
    ISBN 0-230-00125-4 (cloth)
        1. Genetically modified foods. 2. Genetically modified foods–Law and
legislation. 3. Food–Biotechnology. 4. Food–Political aspects. 5. Food law and
legislation. I. Falkner, Robert, 1967-

TP248.65.F66I595 2007
363.19'29--dc22

                                                                2006043716

10   9   8   7   6   5   4   3   2   1
16 15 14 13 12 11 10 09 08 07

Printed and bound in Great Britain by
Antony Rowe Ltd, Chippenham and Eastbourne

To Kishwer and Sophia

Contents

List of Figures and Tables

Acknowledgements

The idea for this book arose at one of the many UN conferences on biosafety that I have attended in recent years, in this case the second Meeting of the Parties to the Cartagena Protocol on Biosafety, held in May–June 2005 in Montreal, Canada. UN conferences can be a frustrating experience, and particularly so when they deal with arcane and technical questions such as the use of identity preservation systems in the identification of genetically modified organisms in international trade. However important these highly specialised debates may be to the functioning of international treaties, they usually fail to illuminate the broader issues that are at stake in international politics. Individuals participating in international conferences are often motivated by more general concerns about environmental protection and international justice, a desire to tame the forces of economic globalization, or an interest in creating a democratic and effective system of international governance. It is this gap between the often-narrow, technical, focus of international policy-making and the broader concerns present in international debates that led me to design this book.

The purpose of this book is to provide an up-to-date analysis of the global GM food controversy, situated within the wider context of international negotiations and domestic politics, environmental protection and development, and trade and international law. When I proposed this idea to Jennifer Nelson a Palgrave, she warmly welcomed it and helped me to get started on this project. Half-way through the gestation of the book, Philippa Grand took over from Jennifer as editor for Palgrave. Her enthusiasm and support for this project were equally fullsome, and thanks are due to both Jennifer and Philippa for making this book possible.

Over several years of researching international biosafety politics, I have incurred personal debts to a great number of colleagues and friends. At Chatham House, where I have carried out biosafety-related research as an associate fellow of the Energy, Environment and Development Programme, Duncan Brack and Richard Tarasofsky have provided a congenial environment for making the worlds of academic research and international policy-making speak to each other. Colleagues in the International Relations Department at the London School of Economics have on many occasions offered critical feedback and encouragement. I have greatly benefited from conversations with Christoph Bail, Duncan Brack, Linda Brown, Aarti Gupta, Les Levidow, Helen Marquard, Ruth Mackenzie, Peter Newell, Richard Tarasofsky, Richard Tapper and Halina Ward. They, and many others, have helped me to clarify my thinking over the years. I thank Carola Kantz for producing the index to this volume. Special thanks are due to Aarti Gupta, with whom I was able to carry out a larger research project on the implementation of the Cartagena Protocol in the developing world, and who has been a constant source of inspiration. Last but not least, I am grateful to the John D. and Catherine T. MacArthur Foundation for providing generous funding for my research on international biosafety politics over the last two years.

The contributors and I are grateful to a number of publishers and journals for permission to reprint material, which appeared elsewhere in an earlier version. Portions of Jennifer Clapp's chapter on 'Transnational Corporate Interests in International Biosafety Negotiations' draw on her previously published article 'Transnational Corporate Interests and Global Environmental Governance: Negotiating Rules for Agricultural Biotechnology and Chemicals', *Environmental Politics* 12(4) (2003), 1–23, reproduced by permission of Taylor & Francis, www.tandf.co.uk. An earlier version of Chapter 3 by Bas Arts and Sandra Mack appeared in *European Environment* 13 (2003), 19–33 © John Wiley & Sons, reproduced with permission. Chapter 5 by Jennifer Clapp is a revised version of a paper published in *Global Governance: A Review of Multilateralism and International Organizations* 11(4) (2005) © by Lynne Rienner Publishers, Inc., used with permission of the publisher. Chapter 6 by Kelly L. Kollman and Aseem Prakash draws on their article 'Biopolitics in the US and the EU: A Race to the Bottom or Convergence to the Top', *International Studies Quarterly*, 47(4) (2003). Chapter 10 by Robert Falkner is a slightly edited version of an article that appeared under another title in *The Pacific Review* 19 (2006).

Notes on the Contributors

Philipp Aerni is a Research Fellow at the World Trade Institute, a research unit in the Center for Comparative and International Studies (CIS) at the Swiss Federal Institute of Technology (ETH Zurich), where he also teaches a course in Science, Technology and Public Policy. He graduated in Geography and Economics from the University of Zurich in 1996 and received his PhD from the Institute of Agricultural Economics at ETH Zurich in 1999. His research interests include international trade regulation, stakeholder attitudes towards agricultural biotechnology, international science and technology policy, development policy and environmental policy.

Bas Arts is a professor at the Forest and Nature Conservation Policy Group at Wageningen University in the Netherlands. His current professional focus is on (1) new modes of governance in international environmental politics (mainly regarding biodiversity, forests and climate change) and (2) the role of private regulation and the power of non-state actors in international environmental politics. He has published several academic articles (among others in *International Political Science Review*, *European Journal of International Relations*, *Policy Sciences* and *Journal of European Public Policy*) and several academic books and chapters (published by Sage, Kluwer, Springer, Edward Elgar and Palgrave). He is co-editor of the series *Non-State Actors in International Law, Governance and Politics* (published by Ashgate).

Thomas Bernauer is Professor of Political Science at ETH Zurich. His research focuses on international and comparative political economy. He is the author of *Genes, Trade and Regulation* (Princeton University Press, 2003), which received the American Political Science Association's Don K. Price Award for the best book on science, technology and environmental politics published in the past 3 years. His research on biotech issues has been published in *European Journal of Political Research*, *International Journal of BioTechnology*, *Journal of Public Policy* and *World Development*.

Jennifer Clapp is a CIGI Chair in International Governance and Associate Professor in the Faculty of Environmental Studies at the University of Waterloo. Her books include *Paths to a Green World: The Political Economy of the Global Environment* (co-authored with Peter Dauvergne, MIT Press, 2005), *Toxic Exports: The Transfer of Hazardous Wastes from Rich to Poor Countries* (Cornell, 2001) and *Adjustment and Agriculture in Africa: Farmers, the State and the World Bank in Guinea* (Macmillan, 1997).

Robert Falkner is Lecturer in International Relations at the London School of Economics and Associate Fellow of the Energy, Environment and Development Programme at Chatham House. He received his PhD in International Relations from the University of Oxford (Nuffield College). His research interests include international political economy, global environmental politics and the role of global firms in world politics. He is co-editor of *The Cartagena Protocol on Biosafety: Reconciling Trade in Biotechnology with Environment and Development* (Earthscan, 2002), and is currently associate editor of the *European Journal of International Relations*. He will be a visiting fellow at the Minda de Gunzburg Center for European Studies, Harvard University during the academic year 2006–07.

Kathryn Hochstetler is Professor of Political Science at the University of New Mexico. Until recently, she was Associate Professor of Political Science at Colorado State University. Her publications include two recent books: *Sovereignty, Democracy, and Global Civil Society* (co-authored, SUNY, 2005) and *Palgrave Advances in International Environmental Politics* (co-edited, Palgrave, 2006).

Grant E. Isaac is the Dean of the College of Commerce at the University of Saskatchewan, Canada. He received his PhD in International Relations from the London School of Economics in 2001. His current research centres on the strategic management of technological innovation with particular emphasis upon intellectual property rights, technological regulation and international trade policy. His recent publications include two books, numerous book chapters as well as articles in leading journals including *Journal of World Trade*, *The World Economy*, *Journal of World Intellectual Property* and *Journal of Applied Corporate Finance*.

William A. Kerr is currently Van Vliet Professor of International Trade at the University of Saskatchewan, Canada and Senior Associate of the Estey Centre for Law and Economics in International Trade in Saskatoon, Canada. He received is PhD in Economics and Agricultural Economics from the University of British Columbia. He has written extensively on biotechnology. He has over 250 academic publications including ten co-authored books. His recent book titles include *Regulating the Liabilities of Agricultural Biotechnology*, *The Economics of Biotechnology* and *Economic Analysis for International Trade Negotiations*. He is currently editor of *The Estey Centre Journal of International Law and Trade Policy*. In 2002 he was made a Fellow of the Canadian Agricultural Economics Society.

Kelly L. Kollman is a Lecturer in the Politics Department at the University of Glasgow. She received her PhD from The George Washington University in 2003. Her articles have appeared in *World Politics*, *Policy Sciences* and

International Studies Quarterly. Her research interests include European politics, environmental policy, transnational social movements and private authority.

Les Levidow is a Senior Research Fellow at the Open University, UK, where his research has focused on the innovation and safety regulation of agricultural biotechnology. This research encompasses the European Union, USA and their trade conflicts. Controversies provide a case study of concepts such as regulatory science, sustainable development, European integration, governance, and organizational learning. His research reports and publication lists are available at the Biotechnology Policy Group website http://technology.open.ac.uk/cts/bpg.htm. He is also editor of the journal *Science as Culture*.

Sandra Mack currently works as a Stakeholder Communications Coordinator at the State Government Victoria, Australia. Before she was an Environmental Policy Officer at the Dutch Ministry of Agriculture, Nature and Food Quality. She holds a MA in the field of Social and Political Sciences of the Environment, and is a graduate of the University of Nijmegen, the Netherlands. She wrote a master thesis on the role environmental NGOs played in the negotiations for the Biosafety Protocol.

Ruth Mackenzie is Principal Research Fellow and Assistant Director of the Centre for International Courts and Tribunals at the Faculty of Laws, University College London. She is co-author of the *Explanatory Guide to the Cartagena Protocol on Biosafety* (IUCN Environmental Policy and Law Series, 2003) and has published numerous articles on international and European regulation of modern biotechnology.

Peter Newell is Senior Research Fellow at the Centre for the Study of Globalization and Regionalisation, University of Warwick. He received his PhD in International Relations from the University of Keele. He is author of *Climate for Change: Non-State Actors and the Global Politics of the Greenhouse* (Cambridge University Press, 2000), co-author of *The Effectiveness of EU Environmental Policy* (Macmillan, 2000) and co-editor of *Development and the Challenge of Globalization* (ITDG, 2002), *The Business of Global Environmental Governance* (MIT Press, 2005) and *Rights, Resources and the Politics of Accountability* (Zed Books, 2006).

Aseem Prakash is Associate Professor of Political Science at the University of Washington-Seattle. He is the author of *Greening the Firm* (Cambridge University Press, 2000), the co-author of *The Voluntary Environmentalists* (Cambridge University Press, 2006), and the co-editor of *Globalization and Governance* (Routledge, 1999), *Coping with Globalization* (Routledge, 2000) and *Responding to Globalization* (Routledge, 2000).

Introduction: The International Politics of Genetically Modified Food

Robert Falkner

Genetic engineering emerged in the 1960s as a revolutionary innovation in biotechnology that some observers expected radically to transform industry and agriculture. As soon as the first genetically modified organisms (GMOs) were field-tested during the 1980s and commercialised in the 1990s, however, genetic engineering became engulfed in a global controversy. Its use in food production, in particular, has provoked highly polarised reactions among producers, consumers, scientists and environmentalists worldwide. While some view it as an essentially beneficial technology that can increase agricultural productivity and help in the fight against malnutrition and poverty, others see it as potentially harmful to humans and the environment. Advocates of the technology urge governments and international organisations to promote its development and commercial adoption by reducing regulatory barriers, while critics demand precautionary regulation to safeguard against potential future harm. Profound uncertainty surrounds the debate about the benefits and risks of genetic engineering, making it even harder to find common ground in this debate.

Over the last decade, the genetically modified (GM) food controversy has become a truly global phenomenon, with environmental and consumer activists coordinating their activities across national boundaries. Public protests and direct action against the planting of GM crops have been reported from around the world. In several European countries, Greenpeace activists have destroyed fields of GM crops and have tried to block the unloading of GM crop shipments from North America. Similar protests have been reported in Brazil and India, where Monsanto and other biotechnology firms have tested and promoted GM crops such as soybeans and cotton. In China, consumer protests have caused several food producers to eliminate GM content from their products, and lawsuits have been filed by concerned citizens to force food manufacturers to label GM content. Several African countries have rejected GM food aid from the United States, despite food shortages, and have vowed not to allow the sale of GM seeds in their markets.

Few other recent technological advances have provoked similar levels of hostility that have spread worldwide with such speed. Passionate debates raged over nuclear energy in the 1970s and 1980s, and localised opposition has emerged more recently against issues ranging from information technology to stem cell research (Bauer, 1995). But few technology debates have 'gone global' to the same extent as the GM food battle. The reasons for this are partly to be found in the recent growth of a transnational political space and the rise of global civil society, in which matters of public concern are now routinely discussed between members of different societies. Anti-GM campaigners have found it comparatively easy to coordinate their activities and exchange information across boundaries. Moreover, the nature of modern food production and distribution has promoted a more global outlook in GM debates. Because GMOs are traded internationally and have permeated the global food chain, the GM debate has inevitably focused on international trade and global governance issues. It should therefore not come as a surprise that transnational anti-GM activism is closely associated with the anti-globalization movement. Not only did both emerge at around the same time, in the late 1990s, but the GM debate also provided anti-globalization campaigners with a fertile ground for arguing that trade liberalisation exacerbates existing inequalities and threatens national autonomy, that transnational corporations have become too powerful, and that international governance is failing to control economic and technological change.

At around the same time, the GM debate started to impact on the international political process. Little-noticed efforts to create international rules on GMO safety (or 'biosafety') had started in the mid-1990s and were expected to be completed in 1999, at a specially convened conference of the parties to the Convention on Biological Diversity (CBD). However, failure to reach agreement at this meeting in Cartagena, Colombia, catapulted the biosafety talks into the limelight of the global trade–environment conflict. The parties continued their search for a compromise and in January 2000 succeeded in adopting the Cartagena Protocol on Biosafety. But despite the fact that this treaty entered into force in September 2003, divisions persist between those countries that demanded strict international biosafety rules and those that feared that the biosafety treaty would impose unnecessary trade barriers and harm the growth prospects of the biotechnology sector.

Two fault lines have characterised the international GMO conflict: one between North and South, and one between North America and Europe. Tensions between developed and developing countries over the question of international biotechnology regulation go back to the 1980s, when developing country representatives for the first time argued for international rules on genetic engineering. They feared that without legally binding international safeguards, developing countries would become the testing ground for what they perceived to be a largely untested Northern technology. At that time, most developed countries opposed these demands, arguing instead for voluntary safety guidelines. Nevertheless, developing countries and transnational activists continued to press Northern governments on this issue and were able to create an international agenda for GMO safety. Despite the successful conclusion of the international talks on the Cartagena Protocol, however, North–South tensions continued to dominate discussion on how to develop further the biosafety regime in areas such as international liability (see Chapter 12, and Falkner and Gupta, 2004).

Somewhat later, in the second half of the 1990s, transatlantic divisions began to emerge that were to play an important role in the international politics of GM food. Growing anti-GM sentiment in Europe forced a change in the European Union's (EU) policy on GMO authorisation and led to a *de facto* moratorium in late 1998 on new GMO approvals and imports. This shift in European policy provoked the first major international trade conflict over GMO safety policies. The world's then leading GMO-producing countries – the United States, Canada and Argentina – threatened to bring a case against the EU under the World Trade Organization's (WTO) dispute settlement procedure, thus raising the diplomatic stakes involved in the parallel efforts to reach an agreement on the biosafety treaty. When the WTO case was finally launched in 2003, the Cartagena Protocol had been agreed, but the GMO-exporting countries unmistakably signalled their intention to fight trade-related biosafety measures that they felt violated international trade rules.

At the time of writing (April 2006), the WTO dispute panel found the EU in breach of WTO rules. Even if this ruling is confirmed after a possible appeal by the EU,[1] it is likely to be a pyrrhic victory for the GMO-exporting countries. First, much of the resistance to GM food in Europe is based on consumer hostility, not regulatory barriers. A WTO ruling that forces the door open to GMO imports from North America is unlikely to convince European consumers and food retailers that GM food is safe. In fact, it may have the opposite effect. Second, the ruling will cement the perception in other parts of the world that biotechnology is being forced upon countries by powerful corporate interests. It is bound to confirm the suspicion in many developing countries that the WTO serves the interests of multinational corporations, not local communities. Third, the GMO dispute may well end up undermining the legitimacy of the WTO if it is seen to erode regulatory autonomy and to ignore the interests of environmental and health protection.

The controversy surrounding the WTO case has highlighted the complex connections that exist between the narrow question of GM food safety and broader concerns about the state of international politics and the global economy. For critics of biotechnology, it is not just about environmental and health risks but also consumer rights, regulatory autonomy and the power of global business. To understand why genetic engineering in food production has become so controversial, we therefore need to take a closer look at the various issues of contention in this debate.

At one level, the GM controversy is primarily about the question of whether genetic engineering poses a risk to human health and the environment. This perspective is at the heart of traditional regulatory approaches, and this is how biotechnology producers and users prefer to frame the debate. Several safety concerns have been raised with regard to the impact of GMOs on the environment and human health: that GMOs might cause a transfer of genes and their traits to wild relatives, which could lead to the creation of 'super-weeds'; that GM plants that contain insect-resistant genes (for example, the Bt gene) would harm not only insect pests but also other species, and thus threaten biological diversity; that GM crops that allow greater use of herbicides and pesticides would have a negative effect on plant and insect life; and that GM food might contain toxic substances or cause allergenetic reaction in the human body. It is important to note, however, that a high degree of uncertainty surrounds these concerns. Until now, there has been no serious environmental damage as a result of GMO releases and no conclusive evidence of harm to human health from the consumption of GM food. In any case, firm conclusions about the safety of GMOs are hard to come by given that long-term threats to biological diversity and ecosystems are difficult to assess. It is not least for this reason that many environmentalists call for precautionary regulation of GMOs, in order to limit long-term and potentially irreversible harm.

At a deeper level, the debate about genetic engineering touches on wider issues relating to the political economy of genetic engineering and regulatory politics. For one, the question has arisen as to how genetic engineering should be regulated. As with other technologies, the choice is between a process-oriented approach that comprehensively regulates all aspects and products of genetic engineering, and a product-oriented approach that focuses only on those GMOs that are suspected of being harmful. The EU has opted for the former, arguing that the novelty and complexity of genetic engineering calls for a comprehensive form of regulation. In contrast, the US bases its regulatory approach on the principle of 'substantial equivalence', which means that GMO products are not considered to be fundamentally different from conventional food products, and are therefore subjected to the same regulatory oversight.

Another, related, question concerns the types of risks and concerns that are taken into account in GMO risk assessment and management. Traditional approaches focus on scientifically demonstrated risks to environment and human health and apply only a limited concept of precaution based on 'sound science'. In contrast, a wider conception of risk and precaution would also consider potential threats from GMOs that are not scientifically proven, or that go beyond strict scientific criteria. Precautionary risk regulation acknowledges the inherent uncertainty involved in scientific debates on GMO risks, and considers risks not only to ecological systems and human health but potentially also to cultural, social and economic systems.

Finally, in regulating genetic engineering, societies need to balance the goals of protection against actual or potential risks and the promotion of scientific and technological progress. Every regulatory system involves some form of cost–benefit analysis, but the long-term risks and benefits from modern biotechnology are difficult to assess, let alone quantify, and cost–benefit calculations will vary from country to country. Countries with higher levels of biological diversity may wish to put greater emphasis on precaution, while countries suffering from widespread malnutrition may value the technological promises of genetic engineering more highly. That GM technology can eradicate malnutrition and even poverty is unlikely, but its potential contribution to raising agricultural productivity and enhancing nutritional values cannot be ruled out. Still, many developing countries are also biodiversity-rich and centres of origin for many important plants, making it even more important for all benefits and risks to be taken into account when deciding on GMO releases and imports.

Given the globalization of biotechnology research, food production and agricultural trade, no country can hope to deal with these complex questions in isolation. International cooperation is needed to create and support regulatory frameworks and to resolve potential conflicts arising from different national approaches. In examining the international dimensions of the GM food conflict, this book contributes to the growing body of research on the challenges of globalization and the prospects for effective global governance. The international politics of GM food provides a vantage point from which to shed new light on important questions about

- the impact that new technologies have on international relations;
- the ability of the states system to create governance institutions that are effective and acceptable to all states;
- the role played by scientists and civil society actors in raising awareness of new risks and promoting normative shifts that inform new regulatory approaches;
- the contribution international institutions make to promoting equity and universal participation in international rule-making in a North–South context; and
- the compatibility of precautionary rules for risky technologies with existing obligations in the field of international trade.

An overview of this book

The twelve contributions to this book analyse the global GM food conflict from the perspective of different academic disciplines and research areas: international relations, comparative politics, trade policy, development studies and international law. The authors examine the political processes that operate at the national, regional and international level and investigate the role played by states, firms and civil society actors. They analyse the transatlantic and North–South dimensions in the global GM dispute and explore its implications for the future of international cooperation, trade and environmental law.

Part I: States, Firms and NGOs in the Creation of the Cartagena Protocol on Biosafety

This book is concerned with the international politics of GM food, and the linkages between domestic politics and international processes. To set the scene, Chapters 1, 2 and 3 review the creation of the international biosafety regime and the role played by different actors and forces: states, non-state actors including corporations and environmental campaign groups, normative factors and political-economic considerations.

In Chapter 1, Robert Falkner analyses the international negotiations that led to the adoption of the Cartagena Protocol on Biosafety in January 2000. He places the negotiations in the context of a wider trend towards international cooperation without the hegemon, or 'non-hegemonic cooperation' (Stiles, 2005). The United States, together with a small group of agricultural export countries, sought first to block, and later to weaken, the proposed biosafety regulations. Yet, as Falkner explains, a broad-based coalition of developing countries and the European Union provided critical leadership during the talks and brought them to a successful conclusion. Against the background of a growing politicisation of genetic engineering and an emerging transatlantic conflict over Europe's GMO import restrictions, a normative shift occurred towards a more precautionary approach to GMO risk assessment and management. The Cartagena Protocol strengthens the regulatory prerogative of GMO-importing countries, and is now being implemented around the world.

In the second chapter, Jennifer Clapp focuses on the role played by transnational biotechnology companies in the international biosafety negotiations. She demonstrates how changes in the underlying business model and market structure in the agrochemical and biotech sectors have driven corporate lobbying strategies in this context. Stagnating profits in traditional agrochemical production and a flurry of merger activities in the chemical and seed sectors increased the stakes involved for those firms investing in GM crop varieties. They were thus keen to ensure that the Cartagena Protocol would not impose burdensome regulations on the bulk commodity trade and were able to maintain a united front throughout the negotiations. Biotech firms exploited the scientific uncertainty surrounding debates on GMO risk to weaken regulatory provisions and relied on the support of the US-led group of agricultural export countries, the so-called Miami Group. Clapp shows that, while corporate lobbying was unable to prevent the creation of an international biosafety regime, it succeeded in watering down key provisions of the final agreement.

The third chapter, by Bas Arts and Sandra Mack, looks at the other major group of non-state actors that played an important role in the biosafety negotiations, namely environmental NGOs. The authors identify the various strategies employed by NGOs to influence the international process – lobby, advocacy, promotion, advice and public pressure – and evaluate their impact on the negotiations. Building on previous work by Arts on the role of non-state actors, their analysis employs a concept of political influence that reflects three indicators: the perception of NGO influence by themselves, which is compared with the perception of other actors and the authors' own analysis of the outcome of the negotiations. The empirical analysis presented in this chapter covers the agenda-setting, negotiation and implementation phase. The authors conclude that lobbying and advocacy roles have provided NGOs with powerful tools to influence the international process, particularly during the creation of an international biosafety agenda. Their influence during the negotiation and implementation phase was more limited, however, as the formal mechanisms of inter-state bargaining and decision-making crowded out activist groups, who depended on like-minded parties to represent their interests.

Part II: Globalization, Corporate Power and International Trade

Moving away from the international biosafety negotiations, the second part of the book focuses on the underlying international political economy of agricultural biotechnology. The two chapters in this part link the international GM food conflict to the globalization of the biotechnology industry, international trade and food aid flows. They transcend a narrow state-centric perspective and move political-economic forces centre stage.

In Chapter 4, Peter Newell examines the role of corporate power in the global spread of GM technology and its impact on developing countries. Through foreign investment and international trade, multinational corporations have come to circumscribe the political autonomy of states in choosing independent biotechnology strategies. Using empirical examples from India and China, Newell shows how biotechnology firms have targeted developing countries in an effort to ensure global acceptance of agricultural biotechnology. But he argues that corporate power has not fully eroded national autonomy. Instead, Newell develops the notion of 'bounded autonomy' to describe the ways in which governments have been able to preserve political space to pursue developmental or commercial objectives of their own. As can be seen in India and China, developing countries can retain a significant degree of autonomy, which is dependent on state strength, their position in global markets, and the role of domestic civil society.

Much of the debate on how to govern transboundary movements of GMOs and the investment activities of biotech firms has focused on commercial transactions. In Chapter 5, Jennifer Clapp draws our attention to a new challenge arising from food aid flows that contain GMOs. In recent years, the United States has repeatedly offered domestically grown GM crops as aid to African countries afflicted by famine. Some countries initially rejected US aid and demanded GM-free food shipments, causing diplomatic tensions with the US and provoking accusations that their anti-GM stance was contributing to hunger in Africa. Clapp investigates the changing political economy of food aid and concludes that economic factors may once again be key motivating factors for giving food aid. Against the background of transatlantic tensions over the EU's GMO moratorium and growing trade barriers against US GM crop exports, food aid is being drawn into the global competition over agricultural market access and GMO safety restrictions.

Part III: The Transatlantic Divide and Its Global Impact

The third part of the book focuses on the transatlantic divide that has emerged over the question of how to regulate GMO safety and trade. While the United States has gradually relaxed the regulatory burden on agricultural biotechnology and become the world's leading GMO exporter, the EU has moved in the opposite direction, creating ever more stringent and comprehensive safety rules. The EU's *de facto* GMO moratorium, which lasted from 1998 to 2004, further aggravated transatlantic tensions and led the US to file a WTO dispute. The three chapters in this part analyse biotechnology policy in the US and in the EU, and examine the impact of the transatlantic GMO conflict on developing countries.

In Chapter 6, Kelly Kollman and Aseem Prakash investigate the domestic sources of the growing transatlantic divide. Using the 'trading up' model in international political economy, they examine the extent to which the EU's stricter regulatory framework has impacted on US policy through transnational economic linkages. Their analysis suggests that a modest 'convergence to the top' has occurred in the US in recent years, with GMO labelling and segregation in agricultural production slowly making their way into domestic biotechnology policy. Crises such as the StarLink scandal have played a critical role in this, as they allowed environmental NGOs to shift the tenor and direction of the domestic debate. Still, the authors conclude that the convergence effect is only a weak one, though it points to broader patterns of regulatory cross-fertilisation across the Atlantic.

Les Levidow shifts the focus to the European Union in Chapter 7. His analysis of the recent shifts in European biotechnology policy points to conflicting policy agendas – trade liberalisation, consumer protection, and civil society participation – that have shaped EU policy. Levidow departs from the common assumption that the transatlantic dispute can be explained with reference to fundamental differences in regulatory philosophy. Instead, the three policy agendas operate across the Atlantic, and give rise to conflict between the US and the EU, as much as between different actors within the EU.

In Chapter 8, Thomas Bernauer and Philipp Aerni also take the transatlantic GMO conflict as their starting point but investigate its impact on the developing world. The authors present a novel interpretation of the roots of the conflict, suggesting that it is less about substantive policy issues than about symbolism and political legitimacy. In their view, Northern pro- and anti-GM interest groups compete for influence in the developing world in an attempt to convince the public that they represent the broader public interest. Given that the economic stakes in extending the conflict to the developing world are comparatively small, they see the focus on the South as an attempt by Northern interest groups to gain public trust. In democracies characterised by pluralist interest group competition, public trust becomes a key political asset particularly in the hands of unelected non-state actors (NGOs, firms). Thus, by claiming to be acting in the interest of a global public good, such as poverty eradication or biodiversity protection, domestic interest groups can increase their discursive power. Bernauer and Aerni further argue that because the battle for public trust is a zero-sum game, the positions in the GMO debate are likely to become more polarised over time.

Part IV: GMO Politics in the Developing World

While Chapter 8 seeks to link the transatlantic GMO conflict with events in the developing world, the two chapters in Part IV take a closer look at the changing nature of GMO politics in developing countries. Chapters 9 and 10 review developments in two regions/countries that are critical to the future of GM crops: Mercosur in South America, which includes the continent's two leading GMO producers, Argentina and Brazil; and China, which has the largest capacity for independent biotechnology research and development in the developing world.

In Chapter 9, Kathryn Hochstetler draws on two analytical approaches to investigate the GMO politics in Mercosur (Argentina, Brazil, Paraguay, Uruguay): multilevel governance and global commodity chains. This dual analytical perspective allows Hochstetler to analyse the diverse range of actors in the political and economic realms that shape GMO policy-making in the region. Among the critical international forces that she identifies are multinational biotech firms, the WTO and the Biosafety Protocol. Her analysis shows that national policy-making in this region takes place under the influence of multiple and overlapping transnational forces. At the same time, however, policy responses have been incoherent and still unpredictable, pointing to the complex and open-ended interplay between divergent national interests, transnational actors and multilateral governance structures.

In Chapter 10, Robert Falkner examines the international forces that have shaped Chinese biotechnology policy. While China is generally considered to possess a high degree of domestic autonomy, Falkner shows how its GMO policy has gradually come under the influence of external factors that have caused a shift towards greater precaution and a slowing down of the GMO approval process. The two most important factors in this context are China's participation in the international biosafety regime, which has strengthened pro-environmental actors within the Chinese state, and growing exposure to international trade flows, which has increased fears among Chinese exporters that GMO restrictions in Europe and East Asia might hurt their export interests if more GMOs are introduced in China. The analysis supports recent arguments that China's international socialisation and economic globalization have led to the gradual internationalisation of domestic policy change.

Part V: The World Trade Organization and International Environmental Law

The final part of this book is reserved for reflections on the place of the emerging global biosafety regime in the wider context of global governance and its relationship with other international treaties and international law.

Chapter 11 by Grant Isaac and William Kerr addresses the complex and controversial question of how the emerging biosafety regime relates to the norms and rules of the World Trade Organization. Their analysis operates at two levels. In the first step, the authors examine the origins, principles and functions of the two regulatory systems of biosafety and trade. These are traced to the different domestic experiences with regulatory success and failure that have fed into the international rule-making process. In the second step, they investigate international efforts to coordinate different and often disparate regulatory systems in a multilateral framework. Isaac and Kerr conclude that while biosafety and trade rules operate in concert to some extent, differences in the principles underlying their regulatory efforts and in the notions of risk regulation employed in them pose a serious challenge to a harmonious relationship between the Biosafety Protocol and the WTO.

In Chapter 12, Ruth Mackenzie rounds off this volume with a review of the contribution made by the Cartagena Protocol to the development of international environmental law. Her analysis highlights key elements that have been controversial during the negotiations and that are likely to have an impact in other areas of environmental law-making. Among these are the precautionary principle, which the Protocol has strengthened through inclusion of precautionary language in its operational part; rules on liability and redress, which have proved more difficult to agree on in the biosafety negotiations, but which are now being elaborated as part of the further evolution of the biosafety regime; and the compliance mechanism, which has, however, failed to move beyond existing models of facilitative compliance. Overall, Mackenzie's analysis suggests that the novel character of biosafety issues has given rise to some important innovations in the biosafety treaty, particularly on precaution and capacity-building. At the same time, however, she warns that the contentious nature of its still unresolved relationship with international trade rules may give rise to legal challenges under WTO law.

Note

1. The WTO panel found that the EU's GMO moratorium had violated WTO rules and communicated its draft decision to the concerned parties in February 2006. After receiving the parties' comments, the panel issued its final decision in May 2006, confirming that the EU had been in breach of international trade rules. The WTO dispute settlement involves a two-step procedure, which gives the EU the right to ask an appellate body to review the panel ruling and issue a final verdict on the case.

Part I States, Firms and NGOs in the Creation of the Cartagena Protocol on Biosafety

1

International Cooperation against the Hegemon: The Cartagena Protocol on Biosafety

Robert Falkner

Introduction

America's retreat from environmental leadership in the early 1990s marks a turning point in environmental diplomacy. Ever since the United States took a backseat role at the 1992 Rio 'Earth Summit', it has repeatedly opposed new environmental treaties that contain binding rules and obligations. Given America's pre-eminent position in the international political economy, the unilateral turn in US foreign policy has come as a blow to ongoing efforts to strengthen global environmental governance. Yet, international environmental policy-making has continued in areas where the US refuses to take on new international commitments as was the case with the Kyoto Protocol, which has entered into force in 2005 despite US withdrawal from the agreement. This chapter investigates another area of international contention, which pitted the US and a small group of agricultural export countries against the large majority of European and developing countries wishing to create internationally binding rules: the Cartagena Protocol on Biosafety to the Convention on Biological Diversity, the world's first international treaty regulating the transboundary movement of genetically modified organisms (GMOs).

The Cartagena Protocol is the result of nearly 4 years of, at times acrimonious, negotiations between GMO-exporting and -importing nations. What started as a relatively unnoticed set of meetings of scientific and regulatory experts in 1996 was soon catapulted into the limelight of the global trade–environment conflict, mainly due to the growing politicisation of agricultural biotechnology in the late 1990s. Developing countries' fears about biotechnology and the European Union's precautionary stance on GMOs in agriculture provided the main impetus for creating stringent international rules that allow importing nations to scrutinise, and potentially reject, international GMO shipments. A small but powerful group of agricultural exporters, led by the United States and comprising Canada,

Argentina, Australia, Chile and Uruguay, opposed these rules but eventually accepted a compromise agreement in 2000 after hard-fought negotiations. Given America's dominance in agricultural biotechnology and opposition to stringent international biosafety rules, the Cartagena Protocol is a remarkable success of non-hegemonic international cooperation.

Non-hegemonic cooperation and international biosafety governance

From the late 1960s to the early 1980s, the United States led international efforts to protect the global environment. It pioneered modern environmental regulation at the domestic level, exported regulatory models and norms worldwide and threw its weight behind the creation of multilateral environmental agreements. But US support for international environmental policy began to wane in the run-up to the 1992 United Nations Conference on Environment and Development (UNCED), and the country has since abandoned its environmental leadership role. It was one of the few countries not to become a party to the Convention on Biological Diversity (CBD, 1992), and although it ratified the UN Framework Convention on Climate Change (1992) it has more recently rejected the Kyoto Protocol (1997) that sets out binding greenhouse gas emission reductions. More often than not, it is now the EU that has moved into a global leadership position on environmental matters (DeSombre, 2005; Falkner, 2005).

One of the puzzling aspects of this shift in US policy is the effect – or to be more precise, the lack of an effect – it has had on international policy-making. For if recent experience is anything to go by, international society has not faltered in its effort to expand the coverage of multilateral environmental policy-making. New and important environmental treaties have come into existence since the early 1990s, with or without US support. And even where the US has taken a more openly hostile stance on international environmental action, other states have not shied away from developing international environmental norms and commitments. In other words, US refusal to participate in these multilateral endeavours may have weakened the resulting governance systems, but has not stopped the process of international environmental regime-building.

A standard assumption in international relations theory has been that lasting international cooperation could only come about with the support of the hegemon. This premise of hegemonic stability theory has come under attack by regime theorists who argue that hegemonic decline does not necessarily spell the end of international cooperation. It may be in the interest of other states to sustain international regimes, and institutionalised cooperation can help reduce the threat of free-riding and defection (Keohane, 1984). But while many regime theorists accept that hegemons play a key role in the creation of regimes, if not in their maintenance, more recent experience suggests that regime-building without – and against – the hegemon has become a reality in international relations (Stiles, 2005).

Recent scholarship on environmental politics suggests that we are indeed witnessing the emergence of a critical mass of countries willing to create governance institutions without the hegemon. On key issues, from climate change to biodiversity and chemical regulation, the EU has filled the leadership gap left by the US and is shaping the direction of international policy debates (Gupta and Grubb, 2000; Burchell and Lightfoot, 2004). This transformation of the EU's role is not complete, however. The EU's performance as a consistent and effective actor varies from issue to issue, and as Vogler (2005: 840–1) points out, coordination problems and conflicts of interest among its 25 member states often reduce the EU's impact in international negotiations. But given the absence of strong environmental leadership from the US, Japan or any other powerful industrialised country, the EU finds itself presented with an opportunity to lead international environmental policy almost by default. This chapter argues that the EU has played a similarly important leadership role in creating a biosafety regime.

But too narrow a focus on state actors and power would miss the important contribution by civil society actors, and ideational forces more generally (Betsill, 2001; Tamiotti and Finger, 2001). This perspective has direct relevance for the biosafety negotiations. Environmental campaign groups played an important role in creating an agenda for biosafety governance, and were supportive of developing countries' efforts to move biosafety issues to the highest level of the international environmental agenda (see Chapter 3). Moreover, rising anti-GM sentiment among European consumers in the late 1990s, which was amplified by transnationally organised environmental campaign groups, played a critical role in transforming the EU's role from being an indifferent and divided actor to exercising critical leadership functions in the final stage of the negotiations.

This chapter examines the factors that have shaped the creation of the Cartagena Protocol. Seen against the background of sustained opposition by the US, the world's leading biotechnology country, and high levels of scientific uncertainty surrounding the GMO safety debate, the creation of the biosafety treaty marks a significant achievement in regime-building without the hegemon.

Background: the international controversy over genetic engineering in agriculture

This section briefly reviews the rise of the global GMO controversy and the emergence of an international biosafety agenda. The subsequent section provides a detailed analysis of the negotiations on the Cartagena Protocol.

The biotechnology revolution in agriculture and rising safety concerns

The use of genetic engineering in agriculture has become one of the main hotspots in international debates on the risks of technology to society and the environment (Bauer, Gaskell and Durant, 2002). Genetic research in the 1960s and 1970s brought about revolutionary changes in plant and animal breeding, allowing scientists to produce genetic change in a more targeted and rapid way, and across species boundaries, through inserting, removing or altering genes. Concerns about the safety of this new technology were first raised in the 1970s. At that time, most countries allowed scientists to regulate themselves and abstained from imposing heavy regulatory burdens. Governments on both sides of the Atlantic were keen to promote what they saw as a promising new industrial sector.

During the 1980s, genetic engineering moved from laboratory experiments to field trials, and the regulatory debate shifted to the risks of deliberate GMO releases into the environment. Biotechnology researchers began experimenting with GM crop varieties that were either tolerant to herbicides (for example, Monsanto's Roundup Ready maize) or resistant to pests (for example, Syngenta's and Monsanto's Bt cotton). The main benefits they sought to produce for farmers were higher and more predictable yields, and thus lower production costs. More recently, researchers have also been working on nutritionally advanced food products, such as vitamin A-enriched rice, though the first generation of commercially introduced GM crops focused primarily on producer, rather than consumer, benefits.

Concerned scientists warned, however, of the risks GMOs posed for the natural environment and human health. Environmental risks include the potential of gene transfer from target plants to non-target plants, leading to the creation of, for example, herbicide-resistant weeds; increased herbicide use in the production of GM crops that are herbicide-tolerant; and negative effects on insect populations, and ultimately bird populations, from the widespread use of insect-resistant GM plants. Human health risks include fears that toxins contained in GM crops could pose a threat to humans; that proteins introduced through genetic engineering may cause allergenetic reactions in humans; and that the use of antibiotic marker genes in plants would lead to human resistance to commonly prescribed antibiotic medicines. Most of these concerns have remained the subject of intense scientific debate, but for many consumers in Europe and elsewhere the risks associated with GM food were real enough for them to reject the technology.

Emerging conflicts over GMO trade

During the 1980s, governments in North America and Europe responded to rising GMO safety concerns with a flexible and deregulatory approach. The United States, the world's leading biotech country, presumed 'substantial equivalence' between conventional and GM food and opted against subjecting all GM products to centralised regulatory oversight. Instead, regulatory authority was divided between agencies already concerned with different aspects of food safety: The United States Department of Agriculture (USDA), the Environmental Protection Agency (EPA) and the Food and Drug Administration (FDA) (Bernauer, 2003: 54–61). Europe seemed initially to follow the US approach. A survey of the regulatory landscape described the situation in most European countries as 'characterized by flexibility and relatively minimal constraints, on the basis of the traditional freedom of scientific inquiry' (Mantegazzini, 1986: 82).

This was to change dramatically, however, with the introduction of a comprehensive, horizontal approach to biotechnology regulation at EU level. Having argued that a patchwork of different regulatory systems in Europe was holding back the consolidation and cross-fertilisation of the European biotech sector, the European Commission succeeded in 1990 to establish an EU-wide regulatory framework. Within the Commission, the Directorate-General Environment managed to gain control over the drafting process and established the EU's GMO regulations as a process-based and precautionary system of risk assessment and management. The choice for a uniform regulatory system reflected a compromise between the desire for a harmonised approach in the interest of free trade in Europe and the need to reassure an increasingly concerned public of the EU's newly acquired regulatory authority in environmental protection and food safety (Pollack and Shaffer, 2005; Chapter 7).

After 1990, regulatory praxis in North American and Europe began to diverge, causing serious friction in international trade and diplomacy. In 1995, the US began the first commercial planting of GM crops. Over the next years, the take-up of GM crops was to rise steadily, making the US the largest GMO producer in the world. By 2000, the year that the Cartagena Protocol was adopted, the US planted GM crops on 30.3 million hectares, accounting for 68 percent of the global GM production area (James, 2000). Inevitably, GM crops also showed up in US farm exports.

The arrival of the first GMO shipments in Europe in 1996 and 1997 led to a rapid increase in awareness of the GMO issue among Europe's consumers. It helped to transform previously isolated and national protests against GM food into a coordinated, European-wide campaign designed to stop the import of GM crops and to block further domestic GMO authorisations in the EU. In a climate of heightened concern about food safety – following the admission by the UK government of a link between bovine spongiform encephalopathy (BSE) in adult cattle and Creutzfeld-Jakob's disease in humans and the imposition of an export ban on British beef – controversy over GMO imports threatened further to undermine public confidence in Europe's regulatory authorities. Reacting nervously to the rise of anti-GM sentiment, some EU member states invoked a safeguard clause in the 1990

regulations, which allowed them to impose national GMO bans. As a consequence, the EU's regulatory approval process came to a halt in late 1998,
and the European Commission was forced into a *de facto* moratorium on
pending and future applications for GMO authorisation (see Chapter 7).
This shift in European GMO policy was to have a major impact on the
dynamics of the international biosafety talks.

The United States accused the EU of protectionism and threatened to take
legal action at the WTO, claiming annual losses of up to $300 million in
farm exports. Frustrated with Europe's regulatory impasse, the US Trade
Representative eventually launched a WTO dispute against the EU in May
2003 (Brack, Falkner and Goll, 2003). Partly in response to the WTO dispute,
the EU introduced in 2004 new rules on GMO labelling and traceability and
reformed its GMO authorisation procedure, which led to the first new GMO
approvals from 2004 onwards. Nevertheless, the EU lost the first round of
the WTO dispute in early 2006.

The WTO dispute not only raised the stakes in the growing transatlantic
conflict over genetic engineering, but also reverberated throughout the
developing world. It heightened the fears of those countries that were keen
to strengthen their right to scrutinise and potentially block GMO imports
through international biosafety rules. During the 1990s, most developing
countries were still in process of developing national biosafety regulations
but lacked the capacity to implement them. They were hoping that an
international biosafety regime would support them through capacity-
building and by providing legal cover against potential challenges under
WTO law. The EU therefore emerged as a critical leader in international
biosafety politics: having developed the most comprehensive, precautionary framework of domestic biosafety regulation, it provided a model that
many developing countries drew on in their own regulatory efforts
(Chapter 10; Paarlberg, 2001). Moreover, by standing up against pressure
from GMO-exporting countries, the EU became a critical ally in the creation of an international biosafety regime. In a sense, the transatlantic
GMO conflict elevated the biosafety negotiations to the status of an anti-
hegemonic struggle in defence of environmental precaution and regulatory sovereignty.

The Cartagena Protocol: from negotiation to implementation

Setting an agenda, agreeing a negotiation mandate

Biosafety issues first emerged on the international agenda in the late 1980s.
In the run-up to the 1992 UNCED, developing countries demanded that the
UNCED agreement addressed the safety concerns surrounding genetic
engineering in agriculture. Their main concern was that the developing
world might become the testing ground for what they perceived to be a

largely untested Northern technology. At that time, both the United States and leading European countries rejected calls for international biotechnology regulation and merely agreed to a provision in the 1992 CBD to consider the need for a biosafety treaty (Zedan, 2002).

Developing countries persisted with their demands for the creation of an international biosafety instrument and were able to build a broad-based and durable coalition of countries from Latin America, Africa and Asia. At the first and second Conference of the Parties to the CBD, in 1994 and 1995, they reiterated their demands for a full-fledged and legally binding treaty, which would incorporate but go beyond the voluntary safety guidelines advocated by developed countries (La Vina, 2002). The fact that the developing world managed to speak with one voice on most of the fundamental demands – though of course considerable disagreements existed on specific regulatory issues – was to play an important role in the successful conclusion of the Cartagena Protocol negotiations.

The other key factor behind the positive outcome of this process was the change in attitude among European countries. Having initially failed to present a united position, the EU gradually came to accept the legitimacy of developing countries' demands, not least since the EU's own 1990 regulation had established many of the provisions that the South demanded at international level. At COP-2 in 1995, the EU eventually adopted a dual-track approach of supporting voluntary guidelines while accepting in principle the start of negotiations on a biosafety protocol (Bail, Decaestecker and Jørgensen, 2002: 168–9). In contrast, the United States viewed such a development with scepticism and continued to challenge the need for binding international rules. As a non-party and observer to the CBD, the US could not, however, prevent COP-2 decisions including the adoption of a biosafety negotiation mandate.

The negotiation process

COP-2 established a Biosafety Working Group (BSWG) that was to meet six times between 1996 and 1999. The BSWG meetings were expected to prepare the ground for adoption of a draft treaty by an extraordinary COP (ExCOP) meeting in 1999. They were attended by governmental delegations of regulatory and scientific experts and were open to observers from non-parties, civil society and the business community (Chapters 2 and 3). Given the novel character of the proposed regulatory framework, and the divergence of views on its content, the first BSWG meeting merely produced a long 'shopping list' (Bail, Decaestecker and Jørgensen, 2002: 172) of regulatory options. Subsequent meetings gradually established a draft text, but without reaching agreement on key elements. Much of the actual negotiation was therefore left until the end of the BSWG process. In fact, the key areas of contention were only resolved after the first ExCOP meeting (in Cartagena, Colombia, February 1999) failed to reach a compromise, and a resumed ExCOP meeting was convened in 2000 to rescue the negotiations.

Despite the continuous presence of environmental NGOs and industry groups, the BSWG meetings attracted little international attention until 1999. Starting in 1998, changes in the domestic political economy of biotechnology, particularly in Europe, were beginning to have a lasting effect on the negotiation dynamic. They would catapult the final phase of negotiations, from 1999 to 2000, into the limelight of the international trade–environment conflict, pitting the US-led coalition of agricultural exporters (the 'Miami Group') on the one hand against the EU and the large group of developing countries ('Like-Minded Group') on the other. Once biotechnology regulation had become politicised in Europe amidst mounting anti-GM sentiment and the EU's *de facto* moratorium on GMO releases, the biosafety negotiations became a key battleground in the struggle between advocates of a free trade WTO order and the interests of environmental protection and regulatory sovereignty. With memories of EU–US differences over the Kyoto Protocol still fresh, the Cartagena Protocol appeared to fit into a broader pattern of transatlantic conflict over the future of global environmental governance.

On most, though not all, contentious issues the US-led Miami Group faced a more or less united front of the EU and the Like-Minded Group of developing countries. The fundamental difference in approach that characterised these two perspectives was that between actual or potential GMO-exporting nations on the one side and GMO-importing nations on the other. Whereas the US and the other members of the Miami Group were most concerned about preventing trade-restrictive regulations, European and developing country negotiators' priority was to strengthen their domestic regulatory capacity and autonomy. By having to comply with environmentally oriented import rules, GMO exporters would have to bear the main costs of the agreement, while importer countries would be its main beneficiaries. This fundamental difference between exporter and importer perspectives shaped the biosafety talks until the end and extended also into the decision-making process of the parties after the Protocol's entry into force.

Core issues of contention

The conflict over specific provisions of the biosafety protocol centred on five issues:

- The scope of international biosafety regulations
- The creation of a regime on liability and redress
- The role of the precautionary principle
- The question of whether, and how, to identify GMOs in agricultural commodity trade
- The relationship between the Protocol and WTO rules

On the question of what *scope* the biosafety protocol should have, the lines of conflict were less clearly drawn than on other issues. From the beginning of the biosafety talks, developing countries were keen to define the scope as broadly as possible, covering all aspects of the application of genetic engineering, including its domestic use, and all products derived from this technology. Most industrialised countries, including the US and the leading European countries, rejected such a wide definition. But whereas the US simply denied the need for a binding international regime, the EU accepted a narrower focus on the transboundary movement of GMOs and their risk to the environment and human health. This issue proved to be contentious throughout the biosafety talks, and resurfaced again at the final round of negotiations in January 2000, when the EU and the US rejected developing country demands for inclusion of GMOs used as pharmaceuticals and products derived from GMOs (referred to as 'products thereof' during the negotiations; see Marquard, 2002: 297–8).

Likewise, both the EU and the US opposed the creation of a binding international regime on liability and redress, a key demand by the Like-Minded Group. Developing countries had argued from the beginning that exporters of GMOs should be held liable if their products, once released into the environment or entering the food chain, caused harm to the environment or human health. They reiterated this demand until the very end of the negotiations, declaring it to be a 'make-or-break' issue. But faced with opposition from the Miami Group and scepticism among leading European countries, the Like-Minded Group had to accept a postponement of a decision on whether or not such a liability regime should be developed. The leading biotechnology countries argued that the conceptual and legal difficulties surrounding liability stood in the way of binding rules. The problems of defining and quantifying environmental harm (for example, to biological diversity), establishing causal links (especially long after the transboundary movement of GMOs has taken place) and identifying those that could be held liable (states or private actors, GMO producers or traders) continue to plague discussions on a future liability regime. But even if these issues could be resolved, the leading biotechnology countries in Europe and North America remain concerned that a binding liability regime would impede biotechnological innovation and international trade.

On the other three core issues, it was the EU and the Like-Minded Group that were largely on the same side, arguing for more stringent biosafety rules than the Miami Group was willing to accept. The precautionary principle proved to be one of the key sticking points in the negotiations. In line with its domestic regulatory framework, the EU argued that importing nations should be allowed to take trade-restrictive measures if GMOs are suspected of causing harm but without conclusive scientific proof. Having initially qualified its support for the precautionary principle with references to 'cost-effectiveness' and the 'reasonable period of time' required for the provision of further scientific evidence, the EU adopted a more hard-line position on precaution after it had lost a WTO dispute over its ban on hormone-treated beef imports from North America (Andrée, 2005: 32–3). Both the EU and the Like-Minded Group felt that the difficulties surrounding research on the long-term effects of GMO releases on biological diversity and human health warranted an approach that was different from the narrow interpretation of precaution in the WTO's Sanitary and Phytosanitary Measures (SPS) Agreement (see Charnovitz, 2002). A wider concept of precaution in the biosafety regime would give importing nations greater leeway in a potential trade conflict with GMO exporters, a point that gained greater importance as US trade diplomats threatened to bring a WTO case against the EU's *de facto* GMO moratorium. The prospect of a future WTO dispute weighed also on the minds of developing country negotiators, who saw the precautionary principle as strengthening their regulatory sovereignty against WTO disciplines.[1]

The thorniest issue in the negotiations proved to be the question of how to treat GMOs used as food or feed, or for processing (also known as agricultural commodities). These GMOs are not intended for release into the environment, and exporter groups argued that they should therefore not be subjected to the same level of regulation as GM seeds. With the growing commercialisation of GM crops in the second half of the 1990s, it became clear that such commodities would make up the vast majority of worldwide movements of GMOs, and precautionary trade restrictions could undermine trade in those products. On the other hand, most developing countries and the EU were keen to ensure that GM commodities were not excluded from the protocol's regulatory provisions. The former argued that in a developing country context, where commodity imports are sold and reused as seeds, the distinction between GMOs intended for release into the environment and GM commodities was untenable. The latter was under domestic pressure to ensure that all GMO trade, including GM commodities imported from the US, was covered by the biosafety treaty (see Council of the European Union, 1999, point 12). Thus, all major negotiating groups saw this issue as critical to their position, and the resolution of this dispute had to wait until the very last moment in the negotiations.

Underlying these disagreements over the content of the biosafety protocol was the fundamental question of how this treaty should relate to WTO rules. At issue was the demand by the Miami Group that the protocol be subordinated to other international agreements, through inclusion of a 'savings clause'. This would guarantee that in a future conflict over GMO trade restrictions, WTO disciplines would apply and the biosafety protocol would need to be interpreted in accordance with WTO rules.[2] Again, both the EU and the Like-Minded Group rejected this demand, arguing that because the biosafety treaty deals with a more specific subject matter under the auspices of the CBD it should take precedence over existing international agreements (see Council of the European Union, 1999, point 15). They were keen to ensure that their use of the protocol's GMO import restrictions could not be challenged in a WTO dispute settlement procedure outside the context of the biosafety protocol. With a looming WTO challenge against the EU's *de facto* moratorium on GMO approvals and US threats against other nations restricting trade in GM commodities, the savings clause became one of the core issues in the final negotiations between 1999 and 2000.

The conflict came to a head at the 1999 ExCOP meeting in Cartagena, which was meant to approve the final text of the agreement. After the Miami Group rejected a compromise proposal by the EU and the Like-Minded Group of developing countries, the meeting collapsed amidst acrimony and mutual recriminations, thus helping to raise the profile of the biosafety talks internationally. Despite their fundamental disagreements, the negotiating parties promised to continue working on a compromise, and a series of informal meetings and the resumed ExCOP meeting in January 2000 laid the ground for the conclusion of the talks. In that final round of negotiations, a compromise was hammered out in bi- or trilateral talks, involving the Miami Group on the one hand, and the EU and/or the Like-Minded Group on the other. During these small-group discussions, Canada came to represent the Miami Group as a whole, but the US acted in the background throughout the talks and played a key role in persuading the Canadian delegation to accept a final compromise on the question of how to identify agricultural commodities.[3] Faced with a united front of European and developing countries, the United States had come to accept the need to reach some compromise, even if unfavourable to its interests. Unlike Canada, which is a party to the CBD, the US knew that as a non-party it would not be able to ratify the biosafety treaty and would thus not be bound directly by its rules and obligations.

The outcome of the biosafety negotiations

The final agreement that was adopted in January 2000 included several compromises between GMO-exporting and -importing nations. While the EU was able to claim victory on key elements of its negotiation position, developing countries celebrated the fact that their demand for an international biosafety treaty had been fulfilled, although they had failed to realise some of their key objectives with regard to its content. Faced with opposition from a broad front of industrialised countries on the question of the protocol's scope and the creation of a liability regime, the Like-Minded Group had to make considerable concessions in these two areas. On the question of the scope of the protocol, the developing countries were unable to have pharmaceutical GMOs and 'products thereof' included in the agreement, but succeeded in a definition of the protocol's scope that was sufficiently broad. The agreement states that it shall apply to the transboundary movement, transit, handling and use of all living modified organisms that may have adverse effects on the conservation and sustainable use of biological diversity, taking also into account risks to human health (Article 4).

No common ground could be found on the question of liability and redress. Faced with US opposition and EU scepticism, the Like-Minded Group reluctantly agreed to include an enabling provision in Article 27 that mandates the Conference of the Parties to 'adopt a process with respect to the appropriate elaboration of international rules and procedures in the field of liability and redress for damage resulting from transboundary movements of living modified organisms (. . .)'. This process has now been started, but it remains uncertain whether any substantial outcome will come out of it.

The Cartagena Protocol's key regulatory mechanism is the advance informed agreement (AIA) procedure, which requires GMO exporters to provide detailed information on the organism in question and to seek the importing nations prior approval before any transboundary movement takes place. Importing nations are to carry out risk assessment before reaching a decision, and in doing so can invoke the precautionary approach. The Cartagena Protocol includes in its operational part language that spells out the circumstances under which such precautionary trade bans can be imposed:

Lack of scientific certainty due to insufficient relevant scientific information and knowledge regarding the extent of the potential adverse effects of a living modified organism on the conservation and sustainable use of biological diversity in the Party of import, taking also into account risks to human health, shall not prevent that Party from taking a decision, as appropriate, with regard to the import of the living modified organism in question as referred to in paragraph 3 above, in order to avoid or minimize such potential adverse effects (Cartagena Protocol, Article 10.6).

While the EU and the Like-Minded Group succeeded in having precautionary language inserted in the protocol, they accepted a compromise on the treatment of commodity trade. The Miami Group succeeded in excluding commodities from the AIA procedure and applying a simplified procedure that would be less trade-intrusive. The procedure creates an obligation on parties to inform other parties of decisions to authorise domestic use of GMOs that may be subject to transboundary movement. This information is to be communicated through the Biosafety Clearing-House, an internet-based information-sharing instrument, enabling importing parties to take a decision on whether or not to permit the import of such commodities.

Critically for exporters, the case-by-case prior approval requirement for every shipment does not automatically apply, unless importing parties decide to subject commodities trade to a domestic AIA procedure (Article 11). The question that proved most difficult in this context, and that nearly derailed the resumed ExCOP meeting in January 2000, was the question of how GM content in commodity shipments is to be identified. The compromise reached in the final hours of the conference stipulates that commodities shipments state that they 'may contain' GMOs (Article 18.2(a)), without specifying the type of GMO and the level of GMO presence in the shipment. This fell well below of the expectations of both the EU and the Like-Minded Group, who wanted to ensure that GMO exporters provided comprehensive information about such exports. But it also was a bitter pill to swallow for the Miami Group, as the parties agreed to specify this provision at a later stage, as happened at the third Meeting of Parties in March 2006.

The question of how the protocol would relate to other international agreements, which had pitted the US-led Miami Group against the majority of other parties, could not be resolved except by inclusion of preambular text that restates the opposing positions while emphasising the mutual supportiveness of the protocol and other agreements:

> Recognizing that trade and environment agreements should be mutually supportive with a view to achieving sustainable development,
>
> Emphasizing that this Protocol shall not be interpreted as implying a change in the rights and obligations of a Party under any existing international agreements,
>
> Understanding that the above recital is not intended to subordinate this Protocol to other international agreement (. . .) (Cartagena Protocol, Preamble).

Thus, by way of a classic diplomatic fudge, the parties defused this contentious issue and left the door open for conflicting interpretations on whether WTO disciplines could potentially overrule protocol provisions.[4] As so often in international negotiations, the parties had to agree to disagree in order to reach an accord.

Implementing the Cartagena Protocol, 2000–2005

Between the adoption of the Cartagena Protocol in January 2000 and its entry into force in September 2003, an Intergovernmental Committee for the Cartagena Protocol (ICCP) met three times to prepare for the first Meeting of the Parties (COP/MOP-1).[5] While the ICCP could not take binding decisions on the development of the protocol, it nevertheless helped to get the Biosafety Clearing-House off the ground and made recommendations on a large list of outstanding issues that future COP/MOPs would have to reach a decision on. The ICCP meetings were characterised by a more amiable atmosphere than the final round of the biosafety negotiations, but little progress was made on key issues such as those relating to GMO identification in trade, liability and compliance. When the COP/MOP therefore met for the first time in February 2004, its agenda was filled with an ambitious programme (Falkner, 2004b; Mackenzie, 2004).

The 50th ratification of the Protocol, which triggered its entry into force in 2003, signalled a dramatic change in the dynamic of the first few COP/MOP meetings. At the time of the first meeting, none of the members of the former Miami Group had become a party to the agreement. Having been relegated to the status of observer countries, their effective participation in the discussions was thus dependent on two factors: the willingness of the COP/MOP to hear their views and let non-parties participate in the proceedings; and the ability or willingness of like-minded parties to represent the views of non-parties.

With regard to participation in the COP/MOP, the United States and other former Miami Group members were entirely in the hands of the parties. As before, they were allowed to make interventions in plenary sessions, working group meetings and also in informal contact group meetings, but the chairs of these meetings reserved the right to give priority to parties, particularly in the end phase of negotiations on critical issues. On several occasions during the Kuala Lumpur conference, Canadian and US delegates complained about the lack of opportunity to adequately voice their perspectives. It had become clear, however, that the parties were intent on pushing ahead with their agenda and were willing to brush aside the objections of those countries that had failed to ratify the agreement. This was less of a problem for the United States, which had entered the biosafety negotiations with the clear expectation of not ratifying the final agreement for as long as domestic objections stood in the way of ratification of the CBD (see Falkner, 2001: 169–72). But this new dynamic posed a more serious constraint for Canada, which had hosted the final round of negotiations in January 2000 and had declared its willingness to ratify, provided that the further evolution of the Protocol would reflect its interests. Now that countries with a predominantly GMO-importing perspective were in charge of the proceedings at COP/MOP-1, non-parties contemplating ratification saw their room for manoeuvre shrinking.

The discussions and decisions at the first two COP/MOP meetings in 2004 and 2005 clearly showed a desire among the countries that had ratified the Protocol to press ahead with the further evolution of the agreement. They listened to non-party perspectives and were keen to ensure that countries planning to ratify in the near future would be included in the deliberations. But when it came to reaching decisions, it was the views of the existing parties that were reflected in the outcomes of both meetings. More sceptical countries such as Argentina and Canada were thus faced with the strategic dilemma of ratifying the Protocol – despite serious reservations – in order to influence future COP/MOP decisions from within, as a party, or to hold back and assess the future evolution of the agreement and its likely impact on their trade interests.

Conclusion: what explains successful regime-building in the case of biosafety?

The Cartagena Protocol on Biosafety has come into existence against US interests and despite US opposition during the negotiations. The United States questioned the need for binding international rules and objected to specific elements of the treaty. It led a small group of agricultural export countries in an effort to limit the trade-restricting effects of the biosafety treaty. As the collapse of the Conference of the Parties in Cartagena in February 1999 showed, the US and its allies were willing to walk away from the negotiations without an agreement if their key demands were not met. In light of the commercial and political stakes involved and the high level of confrontation in the run-up to the final meeting in Montreal in January 2000, it is fair to conclude that reaching an agreement on the biosafety protocol was non-trivial and required considerable diplomatic skill and political leadership on the part of the EU and the Like-Minded Group of developing countries.

Despite the fact that the US is unlikely to become a party, the biosafety protocol is not going to be irrelevant to US interests. As a growing number of countries with major agricultural markets are ratifying the treaty, US agricultural exporters will be faced with import regulations that are based on the Protocol. The treaty will therefore have a *de facto*, if not *de jure*, impact on the US and its farming and biotech sectors. Through bilateral diplomacy and pressure, the US government is trying to ensure that the implementation of the Protocol does not pose an excessive burden on US exporters (Colitt, 2003; Alden, 2003), as happened in 2002 when the US objected to new Chinese safety regulations for GMO imports and extracted major concessions from the Chinese government in the form of interim safety certificates for US GM soybean exports (see Chapter 10). But its economic power and political clout is unlikely to stem the tide of growing GMO import restrictions throughout the world. By legitimising such domestic GMO rules, the Cartagena Protocol has gone a long way to curtail hegemonic power in this policy area.

What factors explain the successful creation of an international biosafety regime, against the apparent interests of the leading biotechnology countries, including the world's most powerful nation, the United States?

First, the nature of the policy area of biosafety and the structure of the underlying international conflict played into the hands of the group of countries that were willing to create an international regime without US support. Unlike other multilateral environmental agreements that seek to reduce pollution at source by phasing out or limiting the production, consumption and transfer of certain pollutants (for example, Montreal Protocol on ozone layer depletion; Kyoto Protocol on climate change), the biosafety regime was designed to strengthen the national prerogative of importing nations and to equip them with the legal and technical means to scrutinise and decide on GMO imports. Even if some countries refuse to join the biosafety accord, those that support and ratify it would still benefit from its existence. They gain international legitimacy for national decision-making and are able to receive capacity-building. Unlike in other treaties where the most important polluters need to be part of the agreement in order to make them work, the Cartagena Protocol can be effective without the support of the leading GMOs producers. Thus, the creation and provision of biosafety governance does not have a strict 'public good' character, and can be achieved with comparatively low costs involved for those supporting the agreement.[6]

Second, it is important to recognise that the Cartagena Protocol was created at the instigation of the developing world. Had there not been a sustained level of demand for international biosafety rules coming from concerned developing countries, it is highly questionable whether the issue of how to deal with GMO safety in trade would have been dealt with in the context of a binding international regime. Most leading biotechnology countries, including the US and the EU, preferred to develop voluntary technical guidelines dealing with safety aspects of genetic engineering, as promoted by the OECD in the 1980s and further developed under the auspices of UNEP in the 1990s. Even though the European Union established the world's most stringent framework for GMO risk assessment and management, which in many ways goes beyond the rules laid down in the Cartagena Protocol, it is questionable whether the EU would have chosen to seek internationalisation of its own rules in the form of a binding regime. European representatives repeatedly emphasised during the biosafety negotiations that the EU did not need internationally binding rules to support its own regulatory framework, but that they were pushing for the adoption of the Protocol in order to help developing countries in their efforts to strengthen domestic risk regulation (Bail, 2000: 23). Thus, without the efforts of the developing countries to create an international biosafety agenda and to keep up efforts to start the negotiation process, a biosafety treaty would most probably not have come into existence.

Third, international leadership played a critical role in countering US opposition to binding biosafety rules. That leadership was provided by the European Union, which changed from an initially sceptical and divided group of countries into a strong proponent of precautionary international biosafety rules. Despite the fact that the developing world had been demanding a biosafety instrument since the late 1980s, it was the emergence of the EU as a leader on GMO safety issues during the mid to late 1990s that transformed the dynamic of international negotiations. EU leadership operated at different levels: by creating the world's most comprehensive and advanced GMO regulations in 1990, the EU provided a regulatory model and contributed to the diffusion of precautionary regulatory practice around the world; by standing up against US pressure to admit GMO imports after the introduction of a *de facto* moratorium in late 1998, the EU showed itself to be willing and capable of countering the hegemon's power, even at the cost of a protracted WTO dispute; and by throwing its weight behind proposals to create a precautionary system of international biosafety regulation and to subject commodities trade to potential import restrictions, the EU closed ranks with the Like-Minded Group of developing countries in their effort to wrestle a compromise agreement from the US-led Miami Group.

Fourth, an explanation of the biosafety negotiations would not be complete without considering the important role of non-state actors, and the rising importance of the precautionary principle in the politics of genetic engineering. Although industry groups had a significant impact on the design of the agreement (Chapter 2), it was the environmental and consumer campaign groups that helped to create the political and normative environment in which an international agreement on biosafety could be concluded. NGOs campaigned for such an outcome at different levels: during the negotiations, where they provided expertise to negotiators, lobbied delegates and exposed the proceedings inside the conference halls to public scrutiny; at the domestic level in the countries that played a critical role during the negotiations, most importantly in the EU where growing anti-GM sentiment had caused a fundamental shift in the EU negotiation position; and generally by promoting awareness of the risks of GM food and highlighting the importance of a precautionary approach to regulating GMOs in agriculture.

Finally, it would also be interesting to reflect on the historical circumstances that shaped the negotiation process, and to ask whether the agreement would have been possible at any other point in recent history. While such conjectural questions are difficult to answer, it may suffice at this point to mention that the late 1990s provided a particularly opportune moment to create a biosafety regime – one that did not exist before and that probably does not exist now. Shortly after the negotiations started in 1996, Europe became engulfed in an increasingly heated debate about the safety of GMOs and GM food. This debate had been going on for some time, in Europe and elsewhere, but the strength of anti-GM sentiment that erupted after 1997/1998 had a clear impact on the biosafety talks. It catapulted the EU into a leadership position and raised the political stakes involved, and provided much-needed background for NGO campaigns in favour of an international biosafety agreement. With the rise of a parallel anti-globalization movement, which contributed in rather spectacular fashion to the collapse of the 1999 WTO Ministerial Conference in Seattle, USA, the late 1990s experienced a sudden eruption of popular unease about international institutions and the legitimacy of trade liberalisation policies. Observers and participants on all sides of the biosafety talks have commented that in the wake of what became known as the 'Seattle debacle', a determination to seek a compromise and avoid a further collapse of international governance efforts was clearly visible at the January 2000 Montreal conference (Bail, Decaestecker and Jørgensen, 2002: 180; Mayr, 2002: 225).

A further special circumstance that should not be forgotten is the fact that the US participated in the biosafety negotiations knowing that it was unlikely ever to ratify the agreement. Not having acceded to the CBD, the 'mother convention' to the Protocol, the US was unable to become a party to the Cartagena Protocol even if it had wished to do so. And since domestic opposition to the CBD has made its ratification by the US impossible during both the Clinton and Bush administrations, US negotiators were cognisant of the fact that whatever biosafety rules came into existence, they would not be directly binding on the US. Of course, the US is now faced with a situation where many of its major export markets are adopting GMO import rules based on Protocol provisions. But it is reasonable to conclude that US resistance to the creation of the biosafety treaty was somewhat muted because of America's observer status at the CBD. This does not, however, render the outcome of the negotiations a trivial case of cooperation. Far from it, the Cartagena Protocol represents an important case of 'non-hegemonic' cooperation. Its global implementation – against US interests – pays testimony to the changing conditions of international cooperation in a world of US unilateralism.

Notes

1. Interview with Indian negotiator, 4 April 2001.
2. In international law, treaties take precedence over other agreements if they deal with a more specific subject matter (*lex specialis*) or have come into existence more recently (*lex posterior*). This can be avoided by inserting a savings clause into a treaty, thus subordinating it to earlier treaties. Article 30(2) of the Vienna Convention on the Law of Treaties (1969) states that '[w]hen a treaty specifies that it is subject to, or that it is not to be considered incompatible with an earlier or later treaty, the provisions of that other treaty prevail'.
3. Interview with US delegation member, 17 July 2001.
4. The continuing divergence of perspectives on the 'relationship' issue in the Protocol can be seen in two contrasting interpretations by members of the EU and US delegations. Afonso (2002) and Safrin (2002b).

5. The Conference of the Parties to the Convention on Biological Diversity is the CBD's governing body. The COP serving as the Meeting of the Parties (MOP) to the Protocol is the Cartagena Protocol's governing body (hence COP/MOP).

6. A UNU/IAS project has calculated that the total amount of international aid spent on capacity-building in developing countries in support of the implementation of the Cartagena Protocol is just over $150 million (2000–05) (see http://www.ias.unu.edu/about/details.cfm/articleID/669 for details of the study).

2
Transnational Corporate Interests in International Biosafety Negotiations

Jennifer Clapp[1]

Corporations played an important role in the negotiation of the Cartagena Protocol on Biosafety (CPB), a global agreement which aims to regulate the trade in genetically modified organisms (GMOs), including bio-engineered seeds and foods. Industry players – individual corporations as well as international and domestic industry associations – strongly resisted strict regulation over the trade of GMOs. I argue here that this position, especially as articulated by the agricultural biotech industry, was conditioned by developments in the broader agrochemical industry. These developments occurred not just in the agricultural biotechnology sector, but also the chemical pesticide sector. The lines which once separated the agricultural chemical sector and agricultural biotechnology sector are becoming increasingly blurred through corporate mergers in an age of economic globalization, and this has influenced industry positions on the global rules that govern agricultural biotechnology.

Recent developments in the biotech and chemicals sectors of the agricultural inputs industry are related to two important factors that influence their profitability. The first relates to the status of intellectual property protection for agricultural pesticides and bio-engineered seeds. The second relates to the evolving state of scientific understanding of the environmental implications of these different agricultural inputs, which affects the extent to which they are regulated by governments. Changes to these two factors occurring over the past 30 years have shifted profit margins in these sectors, and have been important in driving not only industry positions in international environmental treaty negotiations, but also have been a contributing factor to increased corporate mergers and concentration. In this era of change in the global agrochemical and seed industries, industry players have increasingly attempted to influence global economic and environmental policies in ways that help to ensure their own survival and profitability.

Corporate players in global environmental governance

The study of global environmental policy-making in the past 30 years has focused almost exclusively on the role of states in the formation of international environmental treaties. But in the current era of economic globalization, it is increasingly clear that non-state interests have an important role in both the emergence and prevention of global environmental problems and as key players in global environmental governance. This is true not just in the environmental realm, but in international relations more broadly (Rosenau, 1995; Mathews, 1997). Recognizing the limitations of a state-centred approach to understanding global environmental politics, a growing number of scholars in the field are focusing on the role of non-state actors in global environmental negotiations and outcomes.

While the past decade has seen a plethora of studies on NGOs in global environmental politics, there has until recently been much less attention paid to the efforts of business lobby groups and industry representatives to influence the process. The field of international political economy has long included a focus on the structural power of transnational corporations (TNCs) and their resulting indirect influence in global politics (Gill and Law, 1989). There is also an emerging literature on their role as sources of 'private authority' in the global political realm (Cutler, Haufler and Porter, 1999). These players have a strong presence in global environmental negotiations, and have enormous significance for the outcome of international environmental treaties.

Because of the significance of these players, a growing academic interest in the role of transnational corporate groups as global environmental players has emerged in recent years (for example, Finger and Kilcoyne, 1997; Sklair, 2001; Levy and Newell, 2005; Clapp, 2005). This can partly be explained by the growing visibility of these actors in global environmental negotiations since the 1992 Earth Summit (Chatterjee and Finger, 1994). The presence of industry lobby groups at the negotiation of global environmental treaties is now a regular practice, and their involvement in the global politics of environmental issues in general is on the rise. Transnational corporate actors are no longer content to react to international environmental agreement outcomes, but rather are increasingly engaging directly in public debates over global environmental issues. This is particularly true with respect to global environmental issues with clear economic implications for industry, such as the politics surrounding the waste trade, climate change, ozone depletion, chemicals, biodiversity and deforestation, to name some recent examples.

Analyses of the role of transnational corporate interests in these specific cases have only just begun to be undertaken. Some studies have looked at differences in corporate positions among firms in the same sector, such as in the case of energy firms in the Kyoto Protocol talks. These studies seek to explain divergent corporate positions by pointing out differences in social–cultural and domestic political–institutional settings in which these firms operate, especially in cases where firms have similar economic profiles (for example, Skjaerseth and Skodvin, 2001; Rowlands, 2000). In particular, it is frequently pointed out that European firms appear to take a more pro-action line on climate change than North American firms, largely because these settings differ in the two regions, and even among different countries within these regions. This line of enquiry has been important in beginning to unpack the motivations for the lobby positions taken by industry players in international environmental treaty negotiations.

At the same time, though, while not denying the influence that social–cultural and political–institutional factors can have on firm behaviour, Levy and Newell point out that with economic globalization, differences seen in industry positions on key global environmental issues are increasingly dominated by economic considerations. As companies merge on a global scale, the positions taken by industry players on key environmental issues, such as climate change and ozone depletion, are becoming more unified, despite their geographic location (Levy and Newell, 2000: 16). Other studies that have also taken this approach have looked at the ways in which the structural power of these actors in the broader global political economy influences environmental treaty outcomes (Levy, 1997; Levy and Egan, 1998; Newell and Paterson, 1998).

Both of these approaches are important, and they overlap in important ways. Without downplaying the importance of political–institutional and social–cultural factors, and recognizing that they can interface with economic factors in complex ways, I argue that in the Cartagena Protocol, economic considerations are an extremely important explanatory factor for industry's stance. The position taken by industry players in the biosafety case did not divide along regional or country lines. Rather, industry groups presented a fairly unified position in the talks.

Identifying the factors that influence the profitability of firms is key to understanding firms' economic considerations, and thus their positions on international environmental treaties. The status of intellectual property protection on products that are the focus of global negotiations, as well as the state of scientific understanding (which affects domestic as well as international regulation on those products) are strong factors that affect the profitability of the agricultural input industry. The relative profitability of the seed and pesticides segments of the industry in turn affects both the corporate merger activity in these sectors as well as the position they take on environmental negotiations that seek to regulate them. These same factors have had some significance in understanding industry's position in other environmental treaty negotiations. For example, in the case of the Montreal Protocol, it was the development of alternative, patent-protected chemicals, combined with scientific proof of the existence of a hole in the ozone layer that led DuPont to push for an elimination of CFCs (Haas, 1992: 50). Also, some corporations in the fossil-fuel industry tried to exploit scientific uncertainty over global warming as a reason to resist strong rules for curbing climate change (Skjaerseth and Skodvin, 2001: 49; Rowlands, 1995). Though these explanations have relevance in other cases, there has been little academic research that focuses explicitly on them as consistent factors that influence industry's strategy in environmental treaty negotiations.

Corporate influence in the biosafety protocol

Concern over trade in GMOs was the main issue driving the negotiations on the Cartagena Protocol on Biosafety, which began in 1996 and were completed in January 2000 (for background on the negotiations, see Bail, Falkner and Marquard, 2002 and Chapter 1). The main regulatory mechanism ultimately included in the protocol is advance informed agreement (AIA), a version of the prior notification and consent procedure seen in other international treaties. The AIA applies to the import and export of 'living modified organisms' (LMOs). LMOs are genetically engineered, or genetically modified, seeds, as separate from commodities that contain GMOs (Koester, 2001: 82). Negotiations on the protocol were contentious throughout, with controversy over which GMO products were to be covered by the protocol, including whether and how to identify and label GMOs in agricultural commodity shipments, the use of the precautionary principle, and the relationship of the protocol to international trade rules under the World Trade Organization (WTO). Divided over these points, different groups of states emerged at the Cartegena meeting in 1999. The 'Miami Group' (US, Canada, Australia, Chile, Argentina and Uruguay) advocated less stringent regulations, and the 'Like-Minded Group' (including most developing countries) as well as the European Union, advocated more stringent regulations (Newell and Mackenzie, 2000: 313–17).

Industry groups had a particularly strong presence at the negotiation sessions as observers by the end of the negotiations. Their involvement was weak at first, but their interest in participating in the negotiations grew over the course of the talks as it became clear that the treaty would have important implications for them (Reifschneider, 2002: 274). There were eight industry groups represented at the first round of negotiations in 1996, and 20 such groups present at the 1999 meeting in Cartagena.[2] A number of individual corporations sent their own representatives to many of the meetings, such as Monsanto, DuPont and Syngenta (formerly Novartis and Zeneca). Industry organizations with wide memberships were also present by the late 1990s, including the Biotech Industry Organization (BIO), BioteCanada, Japan BioIndustry Organization, the International Chamber of Commerce, the U.S. Grains Council and the International Association of Plant Breeders for the Protection of Plant Varieties.

In the negotiations, the various industry groups participating took similar positions to each other and did not divide sharply along regional lines

(Levy and Newell, 2000: 18). There was a large range of industries involved, from seed producers, to grain traders, to food processors. Though there may have been some differences with respect to how they wanted certain provisions worded, all were strongly opposed to the adoption of strict rules to limit the production and trade in genetically engineered seeds and crops (Sullivan, 2000; BIO, 1999). In particular, industry argued that only LMOs that were intended for release into the environment, such as seeds, should be covered by the treaty, and that commodities containing GMOs, mainly food and animal feed, should not be covered. They argued that to identify and label agricultural commodities would unnecessarily slow international trade (Hogue, 1998a).

Industry groups also argued that reference to the precautionary principle should be limited in the treaty, if not left out entirely. Instead, industry groups strongly argued for the use of risk assessment based on the weight of science as a requirement before countries made decisions on whether to refuse imports of LMOs. Finally, industry lobbied hard that WTO rules should not be overshadowed by the biosafety protocol rules. They wanted to ensure that in cases of conflict, the WTO rules would prevail (BIO, 1999). In taking these positions on the various issues, industry groups tended to favour positions taken by the Miami Group (Hogue, 1998b).

At the final negotiation session in Montreal in January 2000, industry groups had formalized an entity known as the Global Industry Coalition (GIC), chaired by Canada's BioteCanada. The purpose of this group, originally brought together at the 1999 Cartagena meeting, was to give the various industries concerned with the outcome of the protocol one coordinated voice to let the parties know its position. This was seen to be especially important because of a lack of industry coordination at earlier meetings.[3] Because the range of industry groups that had an interest in the biosafety negotiations was much broader than just biotechnology companies, it was felt by some that a network of all of these industries was vital both for the dissemination of information and for coordinating lobby positions.[4] The GIC claims to represent more than 2200 companies based in over 130 countries (Global Industry Coalition, 2000). By forming one industry coalition, the regional division that occurred among the negotiating states, in particular between North America and Europe, did not replicate itself among business actors. According to one participant in these meetings '… the GIC was able to reach full consensus on fundamental positions …' (Reifschneider, 2002: 274).

The final agreement represented a compromise among the parties, and has been characterized as being vague and somewhat ambiguous (Falkner, 2000: 300; Helmuth, 2000: 782). With respect to the question of which GMOs are to be covered by the treaty, in principle it covers all GMOs, but it divides them into categories with different sets of rules applying to each, and some exemptions also apply. LMOs intended for release into the environment (seeds) in the importing country, are subject to a formal AIA procedure for the first international transboundary movement to a country

(Falkner, 2000: 308). Importing countries can reject these if they wish, based on risk assessment. Genetically modified commodities (for food, feed or processing) are exempted from the formal AIA procedure, and instead are subject to identification requirements obliging exporters to state in separate documentation that commodity shipments 'may contain' LMOs, with information about domestic approvals of GMOs to be announced on the internet-based Biosafety Clearing-House. Importers can also reject such shipments, based on risk assessment. In both cases, parties are given the right to make decisions on imports on the basis of precaution in cases where full scientific certainty is lacking. These provisions met with industry's desire to separate out LMOs from commodities containing GMOs, although uncertainty remains about how burdensome the identification rules for LMOs in commodity shipments will turn out in praxis.

The treaty uses the words 'precautionary approach' in the preamble and in the objective, but in the first instance it is mentioned (in the preamble) it specifies that this approach specifically refers to Principle 15 of the Rio Declaration. This compromise is important for industry groups, as Principle 15's version of precaution implies that at least some scientific assessment must be conducted, and that precautionary measures in cases where full scientific certainty is lacking should be 'cost effective'.[5] But the word 'precaution' is not mentioned in the main text. Instead it refers to parties' right to take decisions regarding import of LMOs even in situations where scientific certainty is lacking (Article 10; see also Falkner, 2000: 309). Industry's desire to link the word 'precaution' to Principle 15 of the Rio Declaration was in fact proposed as early as 1997 by the International Chamber of Commerce as a way to incorporate this concept into international environmental treaties (International Chamber of Commerce, 1997).

The vague way in which these conflicts were resolved in the treaty is reflected in the fact that the various negotiating groups all claimed victory when the text was finalized. Industry groups applauded the agreement for the most part (Bruninga, 2000: 145). In particular, they were pleased with the wording with respect to other trade agreements, which in their view obligates parties to apply a scientific risk assessment to back up decisions regarding whether to block the import of a product. Val Giddings of the Biotechnology Industry Organization stated: 'Our rights and obligations under WTO are completely intact' (quoted in Sullivan, 2000: 24). He further stated that 'The Miami group got virtually everything it wanted' (quoted in Helmuth, 2000: 782). Business actors were also pleased with the outcome because they saw that it had several other potential benefits for them. First, the agreement called for a consistent global framework for regulating trade in GMOs, which industry hoped would give a sense of predictability and stability for industry (Bruninga, 2000: 145). The 'may contain' language with respect to identification of agricultural commodity shipments left sufficient room for the grain trading industry to continue its shipments without onerous documentation requirements. Second, the existence of an accord served to 'legitimize' the trade in GMOs (because it calls for regulation as opposed to a ban), which industry saw as important for gaining public trust in GMOs more broadly (Blassnig, 2000: 71).

The agreement came into force on 11 September 2003, 90 days after the 50th party had ratified the agreement. Industry organizations have continued to observe the Meetings of the Parties (MOPs) to the Cartegena Protocol on Biosafety in the years since it was initially adopted. The biotech industry has been less vocal following the adoption of the protocol, while the commodity trade industry has taken a stronger interest. During the final stages of the negotiations and especially since 2000, the commodity trade industry has been actively watching the talks, and has been particularly concerned about developments with respect to documentation and labelling. Grain industry lobby groups, such as the Grain Industry Council, the International Grain Trade Coalition and the U.S. Grains Council, attended the first and second meetings of the parties to the protocol. These groups have continued to lobby for as little information as possible to accompany the shipments of GMOs. The original agreement, however, called for a review and decision regarding the 'may contain' language that is to accompany LMO shipments for direct use as food or feed, or for processing (FFP) within 2 years of the treaty coming into force (that is by 11 September 2005).

The commodity trade industry groups would like to see the 'may contain' language retained in the treaty with respect to LMOs for FFP, in line with the preferences of the major exporting countries of LMOs. At the first Meeting of the Parties (MOP1), in Kuala Lumpur in February 2004, consensus could not be reached on documentation requirements, as certain exporting countries, including Brazil, voiced opposition to any requirements more stringent than the 'may contain' language. It was decided as an interim measure to continue to use the original language on documentation accompanying LMO shipments and an open-ended technical expert group was established to look at identification and documentation for agricultural commodities (Falkner, 2004: 636–37). This expert group, however, was not able to come to any consensus on these issues as exporting countries continued to voice opposition to new requirements, a position that matched that of the grain trade industry groups. At the second Meeting of the Parties, held in May–June 2005, intense negotiations on this issue continued, but also yielded a lack of consensus and the issue is still under review.

Factors affecting biotech industry positions in the biosafety talks

The position of the agricultural biotech corporations on the CPB is a product not only of the developments in the agricultural biotech sector, but also of the developments in the chemical pesticides sector. Indeed, an examination of recent changes in the profitability of genetically modified seeds and agricultural chemicals helps to explain not only industry's positions in the biosafety talks, but also sheds light on the rash of corporate mergers between the two sectors. As a result of this merger activity, three of the top global seed firms, DuPont, Monsanto and Syngenta, are also among the top four global agrochemical firms today. TNC strategy on global environmental treaties such as the Cartegena Protocol on Biosafety is tied to changes in the profitability of these sectors, which in turn is in large part conditioned by both the status of intellectual property protection for these products as well as the state of scientific understanding with respect to health and environmental safety.

Status of intellectual property protection

The status of intellectual property protection affects both the profitability and the negotiating strategies adopted by the agricultural biotech TNCs. Patents for genetically engineered seed have been granted only recently, most of these in the past 5 to 10 years. These patents will thus provide another 10–15 years of intellectual property protection for bio-engineered seeds. The idea of patenting life forms such as seeds is in itself quite novel. It is only since the 1980s that such patents were even allowed in the US and other industrialized countries, and in many developing countries the issue is very controversial. The TRIPS agreement of the WTO aims to harmonize laws for providing intellectual property protection, including patenting of plants and other life forms, across countries, and the biotechnology industry was actively involved in lobbying to ensure that the TRIPS agreement would encourage this. The relative speed with which genetically engineered plants can be developed and brought to market (6 years) compared to conventional plant breeding (10 years) also means a longer time frame for patent protection (Ollinger and Pope, 1995: 56).

The result has been a rush to register patents on bio-engineered crops in the past decade. According to one study, 1370 patents for agricultural biotechnology had been granted by the US Patent and Trademark Office to the top 30 patent assignees by the end of 1998 (ETC Group, 2001: 7). These patents are extremely concentrated in the hands of just a few large agricultural biotechnology companies. Seventy-four percent of the patents for agricultural biotechnology products registered in the US at the end of 1998 were held by just six corporations. This was at a time when the biosafety protocol negotiations were heating up, and the big industry players were keen to influence the outcome so that the life of these new patents was not threatened by any new regulations.

The story on intellectual property protection in the agricultural chemical sector is quite different. Patent protection on many chemical pesticides currently in use has expired. Generic, or off-patent, pesticides account for 53 percent of global pesticide sales, and this figure is expected to rise to 69 percent by 2005 (Kuyek, 2000: 2). Generic brands are generally 25 percent cheaper than brand names (Hartnell, 1996: 380). It takes much more time, and is much more expensive to develop a new pesticide compared to the costs of developing a new genetically engineered seed. This is reflected in the weak growth of the chemical pesticide industry in recent years. In the US the pesticide industry grew only 0.2 percent in 1998, and fell 0.6 percent in 2000 (ETC Group, 2001: 8).

Despite the high cost of developing such chemicals for commercial use, pesticides are still big moneymakers for the agrochemical TNCs and these sales are increasingly tied to the developments in the genetically engineered seeds sector. The global agricultural chemical industry had annual sales in 2004 of approximately US$35 billion versus annual sales in the global seed trade of US$21 billion (ETC Group, 2005a). Though total revenue from seed sales is less than that for agricultural chemicals, the growth in sales of genetically engineered seeds in particular is driving investment in that sector. The biotech seed industry has grown from a value of US$280 million in 1996 to US$4.7 billion in 2004 (ETC Group, 2005a).

The declining patent protection in agricultural chemicals combined with the new patent and other forms of intellectual property protection of genetically engineered seeds has been a driving force in the corporate and sector mergers over the past few decades. Corporations such as DuPont affirm that they are investing heavily in agricultural biotechnology in addition to their chemical operation because they see that the seed sector has great future potential.[6] The resulting corporate concentration in both the agricultural biotech and chemical pesticide industries is striking, and they overlap considerably. Both the chemicals and seeds sectors have become more concentrated over the past decade. In 2001 the top 10 agrochemical companies controlled 84 percent of the global agrochemical market, and the top 10 seed companies controlled 30 percent of the global seed trade and 100 percent of the market for bio-engineered seeds (PANUPS, 2001).

A large part of the rationale for mergers between agricultural chemical and genetically engineered seed companies is that the patents on some of the key agricultural pesticides currently in use have or were due to expire. Most of the genetically engineered seeds brought to market in the past 10 years have been those which are engineered for resistance to chemical pesticides. The sales of these seeds tend to be tied to agreements that require farmers to purchase the seed company's own brand of pesticide, in effect extending the intellectual property protection of the chemical. In other words, the patent protection shifted from being attached to the chemical, to being attached to the seed. In both cases, sales of the chemical are protected. In 2004, 72 percent of newly marketed GM seeds were engineered specifically for herbicide tolerance (James, 2005).

The GM seed industry also has the capacity to ensure that its patents are protected, even in markets where patent laws are not as strictly enforced as in industrialized countries. It has developed genetically engineered genetic use restriction technologies (GURTs), also known as 'terminator' seeds, that produce only sterile second-generation seeds. The use of this technology would prevent farmers from saving seed from their GM crops for future use, forcing them to buy new seeds every year. Various versions of GURTs were patented by Monsanto and Syngenta in the US in the 1990s, but both vowed in 1999 not to commercialize them because of public opposition. In 2000 the Convention on Biological Diversity recommended a moratorium on the use of GURTs and further study on their potential implications. However, a number of agricultural biotech companies have continued research into developing this technology in recent years, and two new patents on it, held by DuPont and Syngenta, were granted in the US in 2001 (ETC Group, 2002). Activist groups are still quite concerned about industry intentions with respect to the use of GURTs (ETC Group, 2005b), while industry has argued that the technology protects biodiversity and removes the need for 'cumbersome administrative procedures, such as those proposed in the framework of the Biosafety Protocol' (International Seed Federation, 2003).

Weight of scientific evidence regarding risks

Profitability of the pesticide and bio-engineered seeds sectors has also been strongly influenced by the availability and general acceptance of scientific studies on their environmental and health impacts. The level of general agreement on the risks associated with these technologies directly affects the degree of domestic and international regulation for approval and use placed on the products in question, and this in turn determines costs for research and development for new products. Higher research and development and regulatory approval costs inevitably eat into profit margins. The degree of scientific 'certainty' over risks posed by chemical pesticides and agricultural biotechnology is not static – it is ever changing as more is learned through study of these technologies. Since the study of agricultural chemical pesticides has been around longer than that for agricultural biotechnology, there is naturally much more known about the risks these chemicals pose than about the risks posed by genetically altered plants.

It is widely accepted in scientific studies that certain agricultural pesticides, particularly those which are persistent organic pollutants (POPs), pose serious environmental and health risks (Colborn, Dumanoski and Myers, 1996). There have been numerous scientific studies on these pesticides, conducted not only by industry researchers but also by academic and government scientists. Though there may be some controversies over the degree of risk and hazard of certain POP chemicals at lower exposures, it is widely acknowledged that they are persistent in the environment, that they bioaccumulate in living organisms and that they cause a number of health problems, especially in high doses (McGinn, 2000; UNEP Chemicals website).

Environmental and health concerns associated with chemical pesticides have contributed to the rising cost for developing and bringing new pesticides to the market. Product development costs in the agricultural chemical industry increased by a factor of four between the mid-1970s and mid-1990s (Hartnell, 1996: 384). The regulatory approval process for a new pesticide is now long and protracted, with environmental regulatory agencies in most industrialized countries examining them extremely closely before approving a new chemical for release. The average cost to develop a new pesticide in the mid-1990s was estimated to be between US$60 million and US$130 million, around a quarter of which was directly associated with regulatory costs (Ollinger and Fernandez-Cornejo, 1995: 16; Hartnell, 1996: 383). This approval process and the associated costs are much higher now than they were when the industry was just getting off the ground in the 1950s. At that time there were fewer regulatory hurdles and the competition was much less fierce. As a result it was much easier then for chemical companies to recoup their research and development costs and still turn a profit.

Development and marketing of new pesticides has slowed considerably since the 1950s because of the more stringent regulatory process (Ollinger and Fernandez-Cornejo, 1998: 142). It typically takes 11 years to bring a new agricultural chemical through the regulatory process, eating into a significant portion of a new pesticide's patent life (Ollinger and Fernandez-Cornejo, 1995: 16). Moreover, today, competition from generic pesticides sales drives down prices for brand names made by leading companies, leaving less profit to invest into research and development of new products. As a result, profitability of chemical pesticides began to decline in the 1980s. At the same time, spending on research and development grew. Worldwide pesticide research and development expenditures expanded 3.2 percent per year between 1983 and 1993 (from US$1.96 billion to US$2.68 billion) while over the same period the global pesticide market grew only 1.5 percent per year (from US$21.74 billion to US$25.28 billion) (Roberts and Begley, 1994: 27). In the US, the proportion of research and development costs that were devoted to getting a product past regulatory hurdles increased from 17.5 percent in 1972 to 47 percent in 1989 (Ollinger and Fernandez-Cornejo, 1998: 140). These high costs to develop new pesticides and the stringent regulatory hurdles were a big reason why corporations sought to engineer plants to match existing, already-approved chemicals. Accompanying this strategy has been a rash of mergers and acquisitions in the agrochemical industry in the past decade. Only a few large companies can afford to put money into research and development of new chemical pesticides, explaining the disappearance of many of the less profitable outfits starting in the 1980s, many of which were acquired by agricultural biotechnology firms.

In contrast to the wealth of scientific knowledge about the risks of chemical pesticides and the extensive regulations that have been put into place, scientific enquiry into the health and environmental impacts of genetically engineered seeds and plants is relatively new. Compared to the number of independent scientific studies on the environmental and health effects of pesticides, there are very few independent scientific studies on the environmental and health impacts of agricultural biotechnology. This lack of a strong body of scientific literature on the environmental and health impacts of genetically engineered seeds and crops has benefited the agricultural biotechnolgoy sector in terms of costs to bring a product to market. The regulatory approval process for genetically modified plants in North America, for example, has been relatively swift when compared to that for new chemical pesticides. The first genetically engineered crops were approved for use (with an initial 5-year trial) in the US in the mid-1990s without much delay. Though the USEPA acknowledged that its initial approval of Bt crops was not based on extensive study of the risks, it re-approved registration of five Bt corn varieties in fall 2001 following a 2-year approval process, for an additional 7 years (though with more stringent requirements than the original approval) (USEPA, 2001; Werner, 2001a). In Canada it takes about 12–18 months for a new genetically engineered seed to go through the regulatory approval process (Doern, 2000: 10). The regulatory process in Europe is more protracted than it is in North America, likely reflecting the stronger affiliation to the hazard-based assessment and precautionary principle in Europe in the face of scientific uncertainty. Though the delays in Europe have been much longer than in North America, the EU has not yet rejected any GM seeds or foods on the grounds that they are unsafe (Levidow and Carr, 2000: 258).

This easier time with registration of new GM crops, at least in North America, affects the cost of bringing these products to the market, making biotechnology a more attractive option for these large agrochemical seed companies than developing new pesticides. Development of new genetically modified crops in the mid-1990s typically took about 6 years and cost US$10 million. This is only one-half the time, and one-fifth to one-seventh the cost of developing a new pesticide (Ollinger and Fernandez-Cornejo, 1995: 17). Biotech companies would undoubtedly like to keep the costs and regulatory approval time down, and have lobbied the US regulatory agencies to this effect, saying that more regulation will only create disincentives to develop new products (Werner, 2001b, 2001c). Industry continues to argue that these technologies are perfectly safe (Bruninga, 2000: 145). The scientific uncertainty surrounding GM crops and the lighter regulatory burden associated with it is likely a key factor in the agricultural input industry's reluctance to accept further regulation on the trade of these products in forums such as the CPB. Indeed, industry used that uncertainty in order to advance its position in the CPB talks (Reifschneider, 2002: 275).

Despite heightened consumer awareness of and concern over potential environmental and health risks associated with GE crops, their use has not declined globally. The area planted with GM crops increased by a factor of over 50 between 1996 and 2005, from 1.7 million hectares to 90 million hectares (James 2005b). But the potential future risk of waning profitability looms over the industry. Growing protests over these crops, particularly in

Europe but also now increasingly in the US and Canada, have put more pressure on government regulatory bodies to scrutinize new products more closely before deciding on whether to approve them. This may ultimately affect the profitability of the agricultural biotechnology. But for now this technology still maintains strong intellectual property protection, and a relatively smooth regulatory approval process, at least in North America.

Conclusion

Industry players took an important role in the negotiation of the Cartegena Protocol on Biosafety. They actively sought to minimize the regulatory procedures incorporated into the Protocol that might restrict the trade in GMOs. This position on the part of industry was in large part a product of economic considerations that were shaped by changing profit margins in the wider agricultural input industry, and in particular the rise in profits linked to the genetically engineered seed sector and the stagnating profits in chemical pesticides sector. The factors that most strongly account for the shifts in economic performance of these sectors are the status of intellectual property protection and the degree of scientific understanding with respect to the risks associated with chemicals and agricultural biotechnology and the regulations that build on this understanding.

A closer look at these factors reveals that the high cost and long time frame for developing new pesticides as the industry faced the expiry of patents on existing chemical pesticides, combined with the swifter pace, lower cost and future patent protection associated with the development of genetically engineered seeds, are important pieces to explaining the strategy of the agricultural biotech industry. A rash of corporate mergers between the chemicals and seed industries over the past decade has facilitated the development of genetically engineered plant varieties that match existing chemicals, ensuring continued profits from both seeds and chemicals. While this strategy provides in effect an extension of the patent life of the chemicals, it also places heightened significance on the need for industry to ensure a continuation of a relatively smooth and easy regulatory process for genetically engineered seeds.

In this context, it is not surprising that corporate players were eager to be involved in the negotiation of the Cartegena Protocol on Biosafety, as the agreement posed a potential threat to industry's long-term strategy. Industry exploited the lack of scientific consensus on the risks associated with GMOs as a means to ensure that the rules were vague and did not interfere with the trade in GMOs, which would enable it to continue to pursue its strategy.

Notes

1.   I would like to thank the Social Science and Humanities Research Council of Canada for financial support for this research project. I would also like to thank

Peter Andrée, Brewster Kneen and Derek Hall for helpful comments, and Barbara Slim, Jeca Glor-Bell and Sam Grey for research assistance.

2. Drawn from attendance lists at these meetings available on the UNEP websites for the CBD and Chemicals Unit.

3. Tom Jacob, DuPont, interview with the author, Washington, D.C., 30 April 2002.

4. Val Giddings, telephone interview with the author, 1 May 2002.

5. Principle 15 of the Rio Declaration reads: 'In order to protect the environment, the precautionary approach shall be widely applied by States according to their capabilities. Where there are threats of serious or irreversible damage, lack of full scientific certainty shall not be sued as a reason for postponing cost-effective measures to prevent environmental degradation' UN, Agenda 21, 1992.

6. Tom Jacob, DuPont, interview with the author, Washington, D.C., 30 April 2002.

3

NGO Strategies and Influence in the Biosafety Arena, 1992–2005

Bas Arts and Sandra Mack

Introduction

Since the adoption of the Convention on Biological Diversity (CBD) in 1992, environmental non-governmental organisations (NGOs) have been trying to influence the negotiation and implementation of a biosafety protocol to the CBD. In trying to impact on the outcomes of the biosafety negotiations, they have used several strategies, such as lobbying, advocacy and pressure. This chapter describes these strategies and assesses whether the NGOs were successful in influencing the formation and implementation of the Cartagena Protocol on Biosafety. In doing so, this chapter contributes to a growing body of literature on the political relevance of NGOs to international environmental politics.

The format of this chapter is as follows. First, a brief discussion of NGOs sheds light on their nature, relations to the UN system, strategies and overall impact on international environmental politics. Second, the concept of political influence as used in this chapter is explained and assessed. Third, the NGOs' strategies to exert influence on the negotiation, decision-making and implementation processes of the Protocol are outlined. Three phases are distinguished: pre-negotiations of the Protocol (1992–95); its negotiation and adoption (1996–2000); and its initial implementation (2000–05). The analysis of each phase is closed off with a short assessment of the NGOs' political influence in that specific phase. Finally, we assess the overall political influence of NGOs, over the entire period, and draw some general conclusions.

NGOs in international environmental politics

An NGO can be defined as a non-profit, non-violent, organised group of people, not established by governments, and not seeking government office (Feld and Jordan, 1983; Willets, 1996). This is a broad definition, making room for churches, scouts, professional associations, business interests and trade unions as well as single-issue organisations such as environmental and human rights groups. It is also a negative definition, only indicating what an NGO is *not*. Following Thomas-Feraru (1974) and Reinalda (1997), and for the sake of this chapter, we would like to preserve the term NGO for those private, non-profit, non-violent pressure groups that pursue certain public aims and that, directly or indirectly, seek to influence political outcomes. Examples are Greenpeace, Oxfam, Amnesty International, WWF and Pax Christi. The question is, however, whether industrial interest groups are covered by this definition or not. In a literal sense they *are*, because organisations such as the International Chamber of Commerce and the World Business Council for Sustainable Development are *themselves* not profit-oriented (although their members are). Moreover, besides pursuing private goals, they also pursue public ones (economic development, sustainability), they are non-violent, and they seek to influence politics in various cases (for example, social policy, WTO regime). Ideologically and functionally, however, these pressure groups are quite different. Therefore in the literature, the distinction is made between NGOs (civic pressure groups), on the one hand, and BINGOs (Business NGOs, or commercial pressure groups), on the other (Chatterjee and Finger, 1994). This article focuses on the first group, NGOs.

The Protocol has been designed and implemented in the context of the United Nations (UN). Therefore, to understand how NGOs could operate in the biosafety arena, it is useful to have a brief look at how the UN and NGOs interrelate. Article 71 of the UN Charter states that the Economic and Social Council (ECOSOC) may consult international NGOs, which are concerned with matters within the ECOSOC's competence, among others UN conferences on special issues (Feld and Jordan, 1983). Since then, it has become standard practice that international NGOs – and national and local NGOs at a later stage (Willets, 1996) – may participate in UN conferences, meetings and law-making, at least if they cover the issues on the agenda and are formally accredited by UN staff. NGOs only have observer status but no formal negotiation role at such meetings, but they may generally make oral statements and disseminate written position papers. In addition, they may have access to working groups in which negotiations between countries take place, which is decided on a case-by-case basis by UN officials (as long as countries do not oppose such NGO access). Finally, countries can include NGO representatives in their delegations at international meetings, but this is dependent on the willingness of individual countries and not common practice yet.

Given these mutual relationships between NGOs and the UN, the former have developed a number of strategies to impact UN decision-making, namely lobby, advocacy, promotion, advice and public pressure (Arts and Mack, 2003; Princen and Finger, 1994; Willets, 1982, 1996). Lobby and advocacy are quite similar. The difference is that lobbying is an *informal* activity and advocating is a *formal* activity. While NGOs may formally – as laid down in the rules of procedure of intergovernmental environmental negotiations – advocate their views and interests within the formal political

process, they may also informally lobby individual government representatives in the corridors or cafeterias of the meetings. Next, promotion includes
activities such as distributing position papers, press releases, scientific findings, draft texts and other information. Related to both lobby and promotion is the strategy of advising country delegations or individual delegates.
Contrary to lobby, however, it is a mutually agreed activity, and contrary to
promotion, it is a rather durable and more direct relationship. Particularly
for developing countries, NGOs often serve as a source of technical, scientific and legal advice. Finally, by mobilising public pressure, NGOs make
their demands known outside the political arena, for example, through campaigns, media, sit-ins or protest demonstrations.

There is much literature and empirical research that show that these
NGOs' strategies are quite successful, since they seem to have become
influential players in international environmental politics. To mention
some examples: NGOs were able to influence the formulation and/or
implementation of (1) the Convention on International Trade of
Endangered Species (CITES), (2) the Antarctic Environmental Protocol, (3)
the Vienna Convention and Montreal Protocol on the Ozone Layer, (4) the
UN Framework Convention on Climate Change, (5) the Convention on
Biological Diversity, (6) Agenda 21, (7) the Convention to Combat Desertification, (8) the NAFTA Supplemental Environmental Agreement, and (9)
legislation concerning the protection of dolphins in tuna fishing (Arts,
1998; Benedick, 1991; Correll, 1999; Chatterjee and Finger, 1994; Dawkins,
1991; Hogenboom, 1998; Princen and Finger, 1994; Ringius, 1997; Risse-
Kappen, 1995; Willets, 1982; Wright, 2000). However, it should be stressed
that these successes refer to elements of these agreements, and that the
majority of treaty texts, policies and measures were still determined by
states, and influenced by other stakeholders.

Conceptualising and assessing political influence

In this article, political influence is defined as 'goal-achievement through one's
own and intentional intervention in a political process' (compare Arts, 1998;
Dahl, 1961; Hubers, 1989). Consequently, NGOs only exercise influence in
the biosafety arena if the following preconditions are met: (1) NGOs achieve
(part of) their policy goals in the formation or implementation of the Biosafety
Protocol; (2) they do so because of their *own* interventions in the biosafety
arena (and not because of third parties, although they may co-operate with
others to achieve goals); and (3) they do so because they have *intentionally*
intervened in this arena to realise these results. Hence, unintentional and
unconscious influences are excluded from the definition, mainly for methodological reasons, because these effects are very hard to measure.

There are several methodologies for assessing political influence, with
the position, reputation and decision methods being the most commonly

known and accepted in political science (Clegg, 1989; Cox and Jacobson, 1973; Dahl, 1961; Goverde et al., 2000; Huberts, 1989; Huberts and Kleinnijenhuis, 1994; Waste, 1986). This chapter builds on the so-called EAR methodology of Arts and Verschuren (1999), which combines elements of the reputation and decision methods. EAR stands for *Ego*-perception, *Alter*-perception and *Researcher's* analysis, whereby the former two elements (EA) employ the reputation method and the latter (R) the decision method.

Methodologically, the reputation method is founded on the assumption that mutual perceptions of influence largely determine social relations (Westerheijden, 1987). Hence, an actor *is* influential when others see him or her as such, and act upon this perception. In other words, the moment NGOs are perceived by 'important others' – such as government representatives, UN officials and independent experts or journalists – as having intentionally influenced the Protocol, NGOs will be *treated* as influential actors by them, which creates opportunities for NGOs to achieve future goals as well. Being a sort of self-fulfilling prophecy, perceptions and 'real' effects are strongly related. Pragmatically, the reputation method can be quite easily applied. A number of in-depth interviews with key respondents suffice, although these respondents should be very well-informed and experienced in the political process under consideration.

To start evaluating the NGOs' political influence in this study, it was determined how NGOs *themselves* (so-called *ego*-perception) and relevant *other* actors (so-called *alter*-perception) perceive the NGOs' influence. Key respondents were asked to give examples of such influence in the biosafety arena (after having explained to them our conception of political influence). In order to be able to conduct the interviews, one of the authors attended the final meeting of the biosafety negotiations in Montreal in January 2000 where the Protocol was adopted. In addition, a number of interviews were conducted by telephone afterwards. In total, 24 respondents were interviewed in the year 2000, some of which were interviewed again in 2005, to assess the impact of NGOs in the implementation phase.

The question is of course whether 'power perceptions' equate to 'real power'. We think only partly so. Therefore we added a third instrument to our method, based on the decision method. This relates to the so-called 'researcher's analysis', which is used as a 'credibility check' of the ego- and alter-perceptions. To that end, the authors analysed NGO documents, UN documents, *Earth Negotiation Bulletins* and relevant academic literature on the Protocol and on the role and influence of NGOs in the biosafety arena. This was done in order to acquire an in-depth view of the decision-making processes in the biosafety arena and the role of NGOs and to assess whether the ego- and alter-perceptions are to be considered credible or not. Where outcomes of all three elements of the EAR methodology coincide, we can derive firm conclusions on the political influence of NGOs.

Pre-negotiations (1992–1995)

We take the adoption of the CBD in 1992 as the starting point of the pre-negotiations. As mentioned elsewhere in this book, Article 19.3 of the CBD urges parties to consider the need for and modalities of a biosafety protocol. Altogether it took governments three full years to finally decide that a legally binding protocol was needed. In 1995, it was agreed to start negotiations on the contents of a protocol.

During this pre-negotiation phase, a relatively small number of NGOs, such as Greenpeace International, Friends of the Earth International and the Third World Network, were present at international biosafety meetings. Although they came from different backgrounds, their common policy goal was to endorse the immediate need for a protocol and to establish a mandate to start negotiations on such an agreement right away. The NGOs' common aim as well as their small number made it easy for them to co-operate, to discuss and to work out common positions at this stage of the process.

The pre-negotiations started with a meeting of Panel 4 – a panel on biosafety of the United Nations Environment Programme (UNEP) – and was followed by two sessions of the Intergovernmental Committee on the Convention on Biological Diversity (ICCBD), a committee that preceded the first Conference of Parties to the Convention on Biological Diversity (COP-1) in the period 1992–94. At those three meetings parties briefly considered the need for a protocol. Panel 4 was open to all kinds of experts, including NGOs. The panel released a report, which made clear that most participants were in favour of a legally binding protocol on biosafety. However, the report was not officially tabled at the first ICCBD meeting, because its status was not clear, while some country groups (particularly the US and its allies which formed the Miami Group later on) opposed the content of the report, and obstructed its tabling. In the end the report *was* introduced and debated, due to G77 and China as well as NGO pressure (Arts, 1998: 216). At the next ICCBD meeting, they continued to urge governments to start negotiations on a protocol immediately, however, without much response.

At the first COP in the Bahamas in 1994 (COP-1), after the CBD had entered into force, NGOs – together with the G77 and China – criticised the fact that the new open-ended ad hoc expert group on bio-safety was asked *again* 'to consider the need for and modalities of a protocol' – after Panel 4 and the two ICCBDs had already done so. In addition, they were propagating a moratorium on the transboundary movements of GMOs until a biosafety protocol would come into being. This plea was more or less ignored, just as NGO proposals on liability, in case GMOs exported by one party caused harm to other parties' biodiversity. During the first COP, NGOs were nonetheless allowed at meetings of the biosafety contact group, which consisted of a small number of government representatives.

Outside the conference hall, NGOs organised their own meetings and workshops, which they used for co-operating and working out interventions. Furthermore, NGOs lobbied delegates in the hallways, during lunch and other possible occasions. In addition, they promoted their views, positions and policy-relevant information by organising forums and by handing out materials to delegates. NGOs particularly worked closely with the group of developing countries (at least those that do not export GMOs), with which their interests were most closely aligned. Generally, many developing countries had only a limited knowledge on biosafety issues because of lack of financial and scientific resources. It was the NGOs which made them aware of the (potentially) negative consequences of the transboundary movement of GMOs for their countries, particularly for their rich biodiversity, traditional agriculture and indigenous people. In that respect, NGO representatives served as legal, scientific and political advisers to developing countries.

The open-ended ad hoc expert group, established at COP-1, first met in Madrid in 1995. The group's task was to elaborate the modalities of a draft biosafety protocol, in preparation for COP-2. The expert meeting was relatively open to NGOs and they made a broad range of contributions. This same group of experts also included NGOs' views in its report (UNEP, 1995). However, NGOs' possibilities for formally intervening became much more limited during COP-2, held also in 1995. One reason was that they were excluded from the proceedings of the biosafety drafting group (*Earth Negotiation Bulletin*, 1996a). Irrespective of continuing resistance from a handful of developed countries, especially the USA, COP-2 nonetheless mandated the negotiation of a legally binding international biosafety protocol. In order to achieve this, an Open-ended Ad Hoc Working Group on Biosafety (BSWG) was established.

In the above, the pre-negotiation phase of the Biosafety Protocol was described in terms of NGO presence and interventions. But were NGOs effective? Did they influence political outcomes? Our EAR method points at mixed results (see also Arts, 1998: 216–20). A number of ego- and alter-perceptions indeed refer to political influence. By passionately advocating the need for a biosafety protocol on the one hand and by co-operating with developing countries on the other, NGOs contributed – according to our respondents – to maintaining the issue on the political agenda. After all, there were strong powers against it (not in the least the USA). Yet it is difficult to fully consider this outcome an NGO success. The need for a protocol was already expressed by different governmental bodies several times (UNEP Panel 4, ICCBDs), but again and again downplayed by the opponents. The 'only' thing NGOs did, together with developing countries, was mobilising counter-power at any time needed. Moreover, NGOs propagated several objectives (for example, a moratorium on the transboundary movement of GMOs), which did hardly get any support in the biosafety arena. In other words, their impact on the *substance* of the agenda-setting process was more or less absent. At the same time, their pressure was 'necessary' to keep the mandatory process for starting the negotiations of a protocol moving ahead. So in terms of building political pressure on delegates in the biosafety arena, NGOs *were* to some extent influential during the pre-negotiation phase.

Negotiation and adoption (1996–2000)

The meetings of this second phase include the six Open-ended Ad Hoc Working Group on Biosafety (BSWG) meetings, the third and fourth Conference of Parties to the Biodiversity Convention (COP-3 and 4), two extraordinary meetings of the COP (ExCOPs), and three informal consultations. The time frame of these meetings ranges from BSWG-1 in 1996 to the final adoption of the Protocol in January 2000 (for an overview see Depledge, 2000; Falkner, 2000; Newell and Mackenzie, 2000; Chapter 1 in this volume).

In July 1996, the BSWG started to elaborate a global protocol on safety in biotechnology. The first meeting was still relatively informal. NGOs were allowed to attend and intervene in all meetings, including the contact groups. Eight NGOs were present at BSWG-1 and formed a coalition. Most governments welcomed the presence of this coalition. As a result, their input was quite intense. In general, NGOs expressed satisfaction with their ability to make interventions. This was in contrast to the closed proceedings of the biosafety contact group which had been held earlier at COP-2, and which excluded observers from attending all-night negotiating sessions on the mandate for a biosafety protocol (*Earth Negotiation Bulletin*, 1996b).

The second and third meeting of the Working Group (BSWG-2 and -3) were also quite informal and open. The number of NGOs present was higher than at the first meeting. At BSWG-2, the Edmonds Institute spoke on behalf of 31 American public-interest groups (*Earth Negotiation Bulletin*, 1997). This time, NGOs were once again allowed to observe all of the meetings, including sub working and contact groups. They were invited to present their views at the start of each negotiation round which concerned a new text proposal. They advocated, *inter alia* (1) the *precautionary* principle as the basis for the Protocol, (2) a *comprehensive* approach (not only GMOs, but all products thereof; and not only including the potential adverse effects on biodiversity, but on human health and indigenous and local communities as well), and (3) strict *liability* for the country of export in case their GMOs, or products thereof, cause damage in the country of import. Besides making known their views and positions, NGOs also offered their scientific and legal expertise to delegates. However, participation in the actual drafting of texts was limited to governmental representatives only.

At BSWG-4 the NGOs' access to the negotiations became much more restricted. The BSWG Bureau decided that NGOs should indeed be allowed to participate as observers, but that they had no right to intervene, negotiate or participate, and that they could be removed from any meeting at the request of any government (*Earth Negotiation Bulletin*, 1998). From that point on, NGO participation in the biosafety negotiations became limited to brief comments at the beginning of formal sessions. In contrast to the NGOs' increasing attendance, their possibilities for intervening at meetings influencing the formal negotiation process had continued to decrease. The Chair of the BSWG explained that it was necessary to maintain such a closed system because NGOs were not supposed to directly influence delegates from other countries during the meetings. Most of the delegates supported the Bureau's decision to restrict NGO access. Naturally, the NGOs were not happy with this imposed restriction.

In February 1999 in Cartagena, Columbia, over 600 participants attended BSWG-6 to finalise a biosafety protocol, which was to be adopted by the so-called 'ExCOP' at the same meeting (only the Conference of Parties can formally adopt a protocol to the CBD). More than 20 NGOs were represented (UNEP, 1999). However, at BSWG-6 as well as at the ExCOP they had even fewer possibilities for intervening than at the earlier BSWGs, and they were no longer allowed to observe all of the meetings. NGO representatives complained about their reduced involvement and influence. As a result, they were not aware of what happened at the meetings and the delegates did not report back to them what had happened in the long negotiation sessions. Despite ten days of non-stop debate, the delegates could not agree on a protocol. This was quite disappointing for the NGOs. The Third World Network (TWN), on behalf of 13 NGOs, stated that the failure of Cartagena bodes ill for the planet (*Earth Negotiation Bulletin*, 1998).

At a following round of three informal consultations, NGOs were again not allowed to observe meetings. This was due to the structure and the confidential nature of the informal consultations, which were devoted to (1) meeting with the chairperson of the negotiations and the spokespersons of the major negotiation groups; (2) consulting within negotiation groups; (3) making informal exchanges between groups; and (4) resolving differences between groups on pending core issues.

In the meantime, with increasing consumer rejection of genetically engineered foods and the legal requirement for segregation and labelling in Europe, the biosafety issue had become embroiled with trade matters (Falkner, 2000). A few weeks before, at the November ministerial meeting of the WTO in Seattle, USA, in 1999, the US and Japan tried to shift the biosafety debate from the UN/CBD, were biosafety was being fought out on its own terms, to the WTO, where trade interests and rules would dominate. Also, Canada proposed the creation of a Working Party on biotechnology under the WTO. However, the biosafety issue became one more strand that broke the Seattle WTO meeting. A number of European environment ministers flew out to Seattle, voicing their objections and openly disagreeing with the European Commissioner for trade, who was prepared to concede to the US-Canada-Japan proposals as part of a trade-off on other issues. Finally, the WTO talks in Seattle collapsed and an attempt to set up a WTO working group on biotechnology failed.

Right after the informal consultations on how to proceed with the Biosafety Protocol, the resumed ExCOP took place in Montreal in January 2000, which was a second attempt, after the failure of Cartagena, to adopt the protocol. At that conference, NGOs had the chance to make a few interventions at plenary meetings, but most consultations were politically charged and took place behind closed doors. NGOs could do nothing, but to await the outcome of the formal meetings. As the days stretched on with little agreement on the outstanding key issues, there was a fear among delegates and observers that again, there would be no protocol at all. Canada and the US continued their efforts to block any real progress in the talks. One of their major points was that they wanted the WTO rules to prevail over the protocol. By the middle of the week core issues were still unresolved and hopes began to be pinned on the arrival of around 40 ministers to pave the way for a political push towards securing agreement on the key issues. Suddenly, after days and nights of tense negotiations, there was a breakthrough. Finally, and almost after another failure to find a compromise, the ExCOP chairman announced at the plenary around five o'clock in the morning that the Protocol was officially adopted.

As the possibilities for formal advocacy in intergovernmental sessions had diminished since WSBG-4, NGOs had to rely on other strategies instead. During the final stages of the negotiations, they were constantly lobbying delegates. In so doing, they had to grasp every chance they could, since they were no longer allowed to participate in the contact and working groups themselves. This 'exclusion' made it more difficult for them to contact delegates personally. Thus, they waited in the hallways of the conference building for delegates who either were coming out or going into the meetings, hoping to share information and views with them. Unfortunately, this was not an easy task. Most of the delegates were tired and exhausted after they had left the long and painstaking meetings, which frequently lasted till late into the night.

Besides lobbying, the NGOs also promoted their views on all of the relevant issues outside the formal negotiations. They continuously handed out information to delegates or deposited it on the tables in the hallways. In doing so, they tried to strengthen their arguments through promoting knowledge about the scientific aspects of biosafety. In particular, some smaller NGOs, such as the Third World Network, the Edmonds Institute, the Institute of Science in Society (ISIS) and the Rural Advancement Foundation International (RAFI) are specialised in doing research and working with scientists. These NGOs regularly released a variety of scientifically based reports and briefings on the biosafety issue. They frequently published *Updates* on scientific findings about the health and environmental effects of GMOs. As a consequence, these organisations were less orientated towards activism than the larger NGOs, such as Greenpeace International and Friends of the Earth International.

Part of the promotion strategy used by the NGOs was to hold informal meetings with negotiating groups or national delegations. In general, most of the delegates had informal contact with NGO representatives on a regular basis. The intensity of these contacts however varied. A distinction needs to be made between representatives from various national governments and negotiation groups. The respondents' common perception was that the developing countries were most open towards NGOs. The main reason which was given to explain this phenomenon was that the stance of the Like-Minded Group was the closest to the NGOs' stance. According to a vast majority of respondents, the NGOs functioned as advisers to the developing countries during the negotiations. On those occasions, NGOs provided the group with policy-relevant information, legal and scientific advice and so on. Some NGOs even supported developing countries financially, so that they were able to send their representatives to the negotiations. In addition, most of the EU delegates, such as those from Germany, Portugal, Denmark and Austria, continuously engaged in both official and non-official dialogues with (mainly national) NGOs. In most cases, the NGOs either corresponded with the relevant ministers, or they had personal consultations with them. Representatives of the Miami Group and especially representatives from the US government were, however, the least receptive to NGO arguments. According to the representatives from US-based NGOs, they were hardly, if at all, able to influence the US administration.

As outlined earlier, NGOs did not attempt to create public pressure in the pre-negotiation and early negotiation phases. In fact, it took quite some time before they actually did so. In Europe, NGOs started to mobilise the public on biosafety issues somewhere around COP-4 in May 1998. It should be noted, however, that this mobilisation process was not only due to the protocol negotiations, but also in response to European discussions on EU-level biosafety regulations. As far as North America was concerned, attempts to mobilise public pressure on the biosafety negotiations themselves did not occur before the resumed ExCOP in Montreal in January 2000.

NGOs' efforts to influence public opinion nonetheless continued to increase. At the two extraordinary meetings – in Cartagena in February 1999, and in Montreal, January 2000 – efforts to mobilise public pressure had reached a climax. NGOs employed several ways for achieving this. One way was to organise demonstrations in order to receive media coverage of their views and to get the public's attention. For example, on 27 January 2000, when the ministers started their meetings in Montreal, Greenpeace activists protested against the position of the Miami Group. They urged the US and Canada to stop obstructing the negotiations and to agree with the majority of countries to set up international rules for controlling GMOs. Activists dressed as butterflies stood behind a wall symbolising US obstructionism and held a banner demanding: 'US and Canada stop blocking Biosafety!' (Greenpeace, 2000). Friends of the Earth International (FoEI) became particularly active when the Canadian Minister of Environmental Affairs, David Anderson, failed to attend the negotiations. When the minister did not appear in Montreal, posters printed in colour with his face on them were hung up everywhere stating: 'Missing Person! Have you seen this man? Wanted in Montreal at crucial biosafety negotiations. Last seen making excuses'. No one knew precisely why, but the minister eventually did decide to attend the ministerial meetings.

In order to mobilise public pressure NGOs used the media as well. They wrote press releases, position papers, briefings and articles on the biosafety negotiations and the issue itself, and these were published in the media and on the Internet. Greenpeace International, for example, released at least twelve press briefings since the Cartagena negotiations started in 1999. In those briefings, Greenpeace explained the context of the negotiations and promoted its view on the protocol. Most of the press releases criticised the Miami Group and other industrialised countries, including the EU.

When dealing with these more activist NGO strategies regarding the biosafety negotiations, it is also important to note that these organisations were very much present in the debate on whether the biosafety issue should be discussed under the UN or WTO umbrella. NGOs and several developing and European countries within the biosafety arena fiercely opposed the latter. Also at the WTO Seattle negotiations (1999), NGOs were actively involved, both inside and outside the Seattle conference centre, and – among others – opposed the dominance of trade over environmental issues (Dale, 2001). In the end, the WTO meeting collapsed, because of civil society opposition and internal divisions between countries. One of the consequences of this was that the issue of biosafety remained firmly within the CBD context.

In the above, the different strategies and activities which NGOs employed during the negotiation phase are dealt with. Again, one may ask whether these were effective. During our investigations, both NGO representatives and other respondents listed a number of issues, which they believed were influenced by NGOs (ego- and alter-perceptions). These perceptions partly coincide. Moreover, we analysed whether these mutually confirmative views were credible, given our knowledge of the policy process (see also Arts and Mack, 2003). As a result, we conclude that the following issues were influenced by NGOs: (1) the adoption of a precautionary approach in the protocol; (2) the (broadened) scope of the Protocol, including risks to human health, agricultural products and commodities; (3) the inclusion of socio-economic considerations – covering potential adverse effects for indigenous and local communities – in the Protocol; (4) the reference to the liability issue (in case GMOs harm the environment and biodiversity of other parties); (5) the handling of the biosafety issue under the auspices of the UN and the CBD (and not under WTO and trade); and, finally, (6) the adoption of the protocol itself, in Montreal in 2000, now by mobilising external public pressure (after the failure of the 1999 negotiations in Cartagena). Of course, these results were not the achievements of NGOs *alone*. They operated in broader coalitions or parallel to like-minded countries. Also, NGOs did only achieve *parts* of their initial objectives. Therefore, the *extent* of their political influence should not be exaggerated (for a more elaborate discussion of this topic, see one of the sections below).

Initial implementation (2000–2005)

The third and final phase includes five meetings: three sessions of the so-called Intergovernmental Committee for the Cartagena Protocol on Biosafety (ICCP-1, 2 and 3) and two meetings of the so-called Conference of Parties to the Convention on Biological Diversity serving as the Meeting of the Parties to the Cartagena Protocol on Biosafety (COP/MOP-1 and 2). The former were organised to bridge the gap between the adoption of the Protocol (January 2000) and its entry into force (December 2003) and to prepare for the implementation of the protocol, which the latter sought to support. At the three ICCPs, many outstanding and controversial issues, relating to the Protocol, were discussed, such as (1) identification and documentation requirements for shipments of agricultural commodities, which (may) contain GMOs, (2) procedures for monitoring of and reporting on (non)compliance by parties to the Protocol, and (3) to start negotiations for a liability and redress mechanism, in case GMOs cause damage in a country of import (UNEP, 2003). However, the ICCPs were just discussion forums, not bodies with formal competencies, as the protocol still had to enter into force. Yet these can be considered important arenas for preparing the implementation process itself.

At all three occasions, several NGOs were present, about 9 in average, including the 'traditional' ones, such as Greenpeace, Friends of the Earth and Third World Network. They could quite freely participate in the discussions. For example, they pointed at the serious impact of GMOs in polluting centres of genetic diversity, as evidenced by the recent accidental contamination of indigenous maize varieties in Mexico. Therefore, they stressed the need for a quick ratification process of the Protocol and for a stringent liability regime. Also, they urged countries to agree upon detailed and separate information and documentation requirements for shipments containing GMOs (or traces thereof) and a strong compliance mechanism (UNEP, 2003).

The Protocol entered into force in September 2003, after the 50th country had ratified it. Soon thereafter, COP/MOP-1 was held in Kuala Lumpur, Malaysia in February 2004. Again the issue of identification and documentation of GMO content in agricultural commodity shipments as well as the issue of liability and redress split the group of countries and inter-ests groups (Falkner, 2004b; UNEP, 2004). Biotech industry and GMO-exporting countries opposed detailed identification and documentation requirements and a strong liability and redress mechanism – as they feared trade restrictions, unworkable documentation requirements and unfair financial claims – whereas most developing countries, the EU and NGOs advocated the opposite, in order to protect the interests of importing countries. The latter group, however, was in a more favourable position, because many of these countries had already ratified the Protocol. In con-trast, most GMO–exporting countries, including the US, had remained non-parties so far. They therefore had no formal negotiation role in Kuala Lumpur (like the NGOs, but of course, these powerful countries can still successfully put pressure on the negotiations). Nonetheless, the propo-nents of the protocol took the opportunity to make some (relatively) strong decisions, given that the opponents were still non-parties. For example, they established an expert working group on liability and redress and a Compliance Committee. At COP/MOP-1, about 12 environmental NGOs were present. They could make a few interventions. At the opening of the Conference for example, Greenpeace, on behalf of several NGOs, drew attention to an incident of contamination in Japan, involving genetically modified canola shipped from Canada. In addition, it made delegates again aware of the NGO agenda on how to elaborate upon and implement the protocol (UNEP, 2004).

COP/MOP-2, held in Montreal, Canada, in June 2005, reproduced all the contradictions of the first meeting, not in the least the identification and doc-umentation controversy (*Earth Negotiation Bulletin*, 2005; UNEP, 2005). The main reason was that, according to the protocol, COP/MOP should now decide on requirements in this field (2 years after the entry into force). Despite this time pressure, parties failed to reach a compromise and postponed a final decision on the issue to COP/MOP-3. This failure was also partly due to the fact that the 'old' coalitions started to shift. An example is Brazil that had developed its own biotech sector the last decade and that had recently moved from one group (the like-minded one) to the other (GMO-exporting coun-tries). This shifting of coalitions made the reaching of a consensus or the formation of a decisive majority even more difficult. NGOs – about 12 were present again – were deeply disappointed about this non-decision on identifi-cation and documentation requirements and felt that the protocol was not respected. On other topics, however, some progress was made in Montreal, for example, an expert group on risk assessment and management was estab-lished. Parallel to COP/MOP-2, the expert group on liability and redress had its first meeting. It discussed several topics quite openly, but no conclusions were drawn so far (as it only needs to deliver a final report in 2007).

From the above it becomes clear that NGOs have continued their interest and participation in the biosafety arena during the initial implementation phase, although the amount of NGOs that were present decreased after Cartagena 1999 and Montreal 2000, where the adoption of the protocol was at stake. But this is a quite normal phenomenon in the policy cycle of international agreements. When negotiations have to be finalised, time pressure increases and an agreement needs to be adopted, the mobilisation of NGOs (and BINGOs) is maximal, definitely so when the agreement itself is controversial and different opposing interests have to be settled. After the agreement has been adopted, however, the ratification process starts, during which the agreement itself has not a binding legal status. Such a period is less interesting for NGOs, as the substance of the agreement itself is less on the agenda. Yet future parties will start – although informally – to prepare themselves for the implementation process at intergovernmental meetings. Therefore the core of NGOs remains involved during the ratification process. After the entry into force, though, things become more interesting again and NGOs will generally increase their attention. This is also the case for the biosafety arena: we observe a small increase from about 9 environmental NGOs involved during the ratification process (ICCBs) to about 12 ones after the entry into force of the Protocol (COP/MOPs). They interfered – or at least tried to do so – in all the implementation issues on the agenda, but mainly on the identification and documentation requirements with regard to shipments, the compliance mechanism and the topic of liability and redress. They again deployed the strategies of lobby, advocacy, promotion and advice to achieve their goals. Besides, NGOs continued to create public pressure. For example, Greenpeace protested against the undermining of the protocol by WTO rules and the Bush Administration for suing the European Union, because it has restricted import of GMOs.

According to the respondents, NGOs influenced the topics of identification and documentation, liability and redress as well as compliance during the implementation phase so far. However, these perceptions cannot be confirmed by our own analysis. First, the issue of documentation has remained unresolved so far, so one can hardly speak of any meaningful political outcome up till now (June 2005). Second, a Compliance Committee was indeed established, but was probably done so *anyway*, irrespective of the NGOs' presence and interventions, given common practices during first meetings of COPs of international environmental agreements. These always start with establishing secretariats and subsidiary bodies, including one related to compliance (in nearly all cases). Third, an expert working group on liability and redress was established as well, but that fact cannot be contributed to the role of NGOs either. This outcome was a logical one given the Protocol text in Article 27, which states that the first COP/MOP should start a process to elaborate upon a liability mechanism. To install a working group is then the most normal (and most minimal) thing to do.

NGOs' political influence

In the previous sections we showed how NGOs have intensively tried to influence the biosafety negotiations in their interest and whether they were able to affect certain subjects and decisions. Here we will elaborate on this question of political influence by giving an *overview* of all three policy phases and by further discussing the *extent* of influence. All in all, we conclude that NGOs *did* exert influence on agenda-setting and design of the Biosafety Protocol, whereas such influence was absent during the implementation phase so far. More specifically, NGOs contributed to achieving the following outcomes:

1. maintaining the need for a protocol on biosafety on the political agenda during the pre-negotiation phase (as there were strong powers against it);
2. the adoption of a precautionary approach in the Protocol (Preamble and Articles 1 and 10);
3. the (broadened) scope of the Protocol, including risks to human health and covering agricultural products and commodities as well (Article 4);
4. the inclusion of socio-economic considerations – including potential adverse effects for indigenous and local communities – in decision-making on GMO import (Article 26);
5. the reference to the liability issue (in case GMOs harm the environment and biodiversity of other parties) (Article 27);
6. the mobilisation of pressure on governments, inside and outside the biosafety arena, to keep the biosafety issue under the auspices of the UN and the CBD (and not under WTO and trade); and
7. the mobilisation of pressure on governments to finally adopt the protocol in Montreal in 2000 (after the failure of the 1999 negotiations in Cartagena);

It is important, however, to consider the limitations of NGO influence and power. In our definition of 'influence', actors have political influence when they are able to achieve their goals based on their own interventions in the international process. This is a demanding definition, and in most cases NGOs will have only *some* influence at best. Indeed, as the biosafety case suggests, NGO influence was rather low when measured in terms of goal achievement and direct NGO interventions.

As far as goal achievement is concerned, one should take into account the fact that nearly all these outcomes have been compromises between the different countries and the different interest groups, so that NGO proposals, for as far as these (seem to) appear in formal policy texts, are strongly watered down. To illustrate this point, Clapp (Chapter 2) analyses how the biotech industry – generally opposing the NGOs' views – was able to achieve results as well. Its involvement grew rapidly during the negotiations, with, for example, 8 industry groups present in 1996 and more than 20 in 1999.

In addition, individual transnational corporations, such as Monsanto and DuPont, closely followed and addressed the negotiations. Industry was also able to form one umbrella organisation, the Global Industry Coalition. To further its interests, it closely co-operated with the 'Miami Group'. In the end, this coalition of biotech industry and GMO-exporting countries was able to water down many of the initial proposals. For example, GMOs were re-defined as LMOs, text elements that seem to make Protocol regulations subject to WTO rules were also included in the protocol, the precautionary approach was linked to scientific risk assessment (hence, we deal with a 'soft' version of the precautionary principle here) and, finally, a liability regime was omitted from the Protocol.

Next, as far as 'own interventions' is concerned, one should consider the fact that NGOs have achieved (some) results in co-operation with or parallel to other parties, such as developing countries of the 'Like-Minded Group' and several EU members. Hence, NGOs have been part of a broader coalition, so that successes have never been the result of their interventions alone. Besides, within this coalition and compared to their governmental allies, NGOs possess a weak position, since they do not have any formal negotiation role in UN law-making. Formally, they only possess an advocating role. But even this role is limited, since the extent to which it can be played heavily depends on UN personnel and policy phase. Whether NGOs can participate in smaller contact and working groups is decided by secretariats on a case-by-case basis. Moreover, the more negotiations reach the final stages, the more NGOs are excluded from the process. Therefore it is safe to conclude – given their limited possibilities to formally intervene – that the NGOs' *direct* impact on formal decision-making is marginal. Of course, NGOs have compensated this limited formal access by applying other strategies, as was shown in the above, and have definitely increased their political influence as a consequence.

Our data not only make it possible to draw conclusions on the NGOs' political influence, but on the effectiveness of NGO strategies as well. The conclusions drawn so far show that NGOs had the most influence *outside* the formal negotiation process. As a result, the advocacy strategy – linked to the formal negotiation process – can be considered rather insignificant. Outside of the formal negotiations, the strategies of public pressure, promotion, lobby and advice were of greater importance. Most respondents perceived the latter two strategies as the most effective ones, especially with regard to maintaining personal contact with government representatives and providing scientifically based information to developing countries.

Conclusion

We would like to finalise this chapter by drawing three main conclusions. First, environmental NGOs, such as Greenpeace, Friends of the Earth, Third World Network, RAFI and others, employed a number of strategies in order to influence the formation, contents and (initial) implementation of the Cartagena Protocol on Biosafety in the period 1992–2005: lobby, advocacy, promotion, advice and public pressure. Of these strategies, lobby and advice turned out to be the most effective. Second, by employing these strategies, the NGOs were able to influence (1) the *agenda-setting* and *adoption* process of the protocol (ince the pressure of NGOs contributed to mandating the start of the negotiation process of the Protocol in 1995 as well as to adopting the final compromise in 2000), and (2) the *contents* of the Protocol (notably its legal status, being linked to the CBD, the inclusion of the precautionary approach, the broadening of the Protocol's scope, the inclusion of socio-economic considerations and the liability issue). However, the implementation phase has not (seriously) been affected by NGOs so far, although one should take into consideration that this phase has just started and has not produced many meaningful political outcomes so far. Third, the *extent* of the NGOs' influence is – overall – limited, as they have no formal negotiation role to play in UN decision-making, so that their impact in the biosafety arena was dependent on support by like-minded countries and stakeholders. Moreover, many of the original NGO views and proposals, for as far as these were taken over or endorsed by governments, were strongly watered down during the negotiation process.

Part II Globalization, Corporate Power and International Trade

4
Corporate Power and 'Bounded Autonomy' in the Global Politics of Biotechnology

Peter Newell

Introduction: the global politics of biotechnology

The politics of biotechnology are increasingly played out against a background of conflicts over trade, aid and development. Processes of democratic state decision-making regarding technology choice and agricultural futures are increasingly subject to global and private commercial scrutiny. This is significant both for the state autonomy it erodes and for the possibilities of a biotechnology policy which might be considered responsive to the needs and priorities of poorer groups.

Understanding the forms of power which are manifested in these developments requires us to connect questions of corporate strategy with global policy processes in which corporations are privileged actors. The role of leading biotechnology firms in technology development, control and in shaping the regulations that have evolved for managing technologies is examined. The specific context and challenge for developing countries is then introduced with particular reference to the globally significant cases of India and China. These are drawn upon, alongside other examples, as a basis for exploring the micro-politics of corporate power and the ways in which these enable and constrain a particular politics of biotechnology, a particular type of development and a particular type of democracy.

Corporations in the governance of biotechnology

However much exaggerated the potential has been, biotechnology has come to be regarded as an important tool by which developing country governments can boost economic growth and combat problems of chronic food insecurity (Nuffield Council on Bioethics, 1999; Lipton, 1999; Paarlberg, 2001; Thomson, 2002). Because of the expertise, economic power and influential policy networks in which biotechnology firms are nested, they have come to play a key role in shaping regulatory developments at the national and international level (Glover and Newell, 2004). At the level of implementation, firms act as 'street-level bureaucrats' (Lipskey, 1980) overseeing compliance with biotechnology and biosafety regulations.

The key role that firms play in systems of biotechnology governance the world over, but perhaps especially in the developing world, should not be confused, however, with more general claims about the impotence of developing countries in the face of the undoubted might of leading biotechnology corporations (Action Aid, 2003; Christian Aid, 2000; Brac de la Perriere and Seuret, 2000). Developing countries are often presented as victims of unscupulous multinational countries seeking to exploit their economic vulnerability and lack of political capacity in order to test and develop GM products that have been rejected in the North. It is assumed, therefore, that developing countries are more strongly influenced by international political and economic pressures that constrain and enable particular policy options and processes, than is the case for many Northern countries.

It is certainly the case that the reality of power imbalances between large biotech firms and developing country governments, combined with many governments' pressing need to attract foreign direct investment, means that it may be harder for developing countries to prioritise food security and poverty concerns over the imperative of attracting capital. Aside from broader pressures to attract FDI from actors such as the World Bank and IMF, regulators in developing countries have been subject to intense industry pressure to speed up application procedures for biotech developments to avoid 'undue delay' Evidence is provided below of aggressive bilateral lobbying through trade officials, the strategic use of aid and intense commercial pressures applied on developing countries by leading biotech firms. But such pressures have been applied to developed countries too, amid intense lobbying to get GM crops rapidly approved in lucrative markets and ongoing threats to bring trade action against the European Union by the US government on behalf of its biotech firms. The key issue then is the differential ability of states to withstand those pressures, should they wish to, to exercise autonomy in their decision-making about the role of biotechnology in their development.

This chapter explores the extent to which and the ways in which state autonomy regarding key decisions on the future of biotechnology in agriculture is circumscribed by the political power and corporate strategies adopted by leading biotech firms. It introduces the notion of 'bounded autonomy' to describe the ways in which governments have been able to preserve space to pursue developmental or commercial objectives of their own, often against the expressed wishes of biotechnology companies and those governments that act to advance their interests. The extent of this autonomy is seen to be a function of state strength, market positionality and the role of civil society which combine in unique ways in diverse settings.

Bounded autonomy

Given what is at stake in the global debate about biotechnology, it is unsurprising that so many powerful forces have been brought to bear upon governments that are expected to balance the risks and opportunities associated with the technology in their decision-making about its role in their national development. From multinational companies to vocal civil society activists and well-organised scientific communities, competing interests are aligned to press governments to adopt their positions. The extent to which countries are able to resist such pressures, to preserve their sovereignty, depends very much on the state in question.

'Bounded autonomy' describes the policy space that governments are able to preserve in the face of these pressures in which they can formulate and implement national development strategies. Regarding biotechnology, the issue is the ability of governments to exercise political and social control over the technology in ways which serve broader developmental ends. It resonates in this sense with Keeley's (2003a) use of the developmental state concept in relation to China's strongly state-managed biotechnology trajectory, as well as Evans' (1995) notion of 'embedded autonomy' applied to understanding the role of the state in industrial policy. In the latter case, however, the focus is on the capacity and autonomy of the bureaucracy and the extent to which its relationship to productive classes (industrial capital) *within* the country is sufficiently 'embedded' that they are able to mobilise industry towards the fulfilment of broader developmental projects. The use of the term 'bounded' is intended to describe the constraints on the exercise of autonomy, rather than to imply that the state acts in such situations as a bounded entity. The analysis below demonstrates quite clearly the porous nature of the boundaries of state decision-making where some parts of the state are more internationalised than others and global actors are present within the state, breaking down distinctions between domestic/international and public/private. There are a number of factors at work here.

The nature of state–society relations is central to appreciating the extent to which the state is insulated from civil society pressures, where in China the early and aggressive development of GM crops was attributed, at least in part, to the absence of activist pressure (Paarlberg, 2001; Newell, 2003b). Likewise, Paarlberg (2001) relates the more precautionary approaches to biotechnology adopted by governments such as India and Brazil to the meddling of NGOs, particularly Western-based activists, in decision-making about biotechnology. In doing so, however, he overlooks the many other sources of domestic resistance and legitimate concern regarding the scope and pace of the technology's development discussed below.

Though international law can, on occasion, act as buffer between external pressure and domestic policy autonomy, as a guide to formulating a coherent biotechnology policy, the international instruments that 'govern' the technology are of little use. Most countries are faced with inconsistent and mixed messages from international organisations active in the biotechnology area, which place different emphasis on the balance between trade, environmental protection and food security in the design of regulations. These include the Cartagena Protocol on Biosafety and the WTO agreements on standards (SPS and TBT), agriculture (AoA) and intellectual property rights (TRIPS). Ambiguities and tensions among these agreements could translate into a political opportunity for countries to adopt a variety of national political strategies regarding the regulation of biotechnology products, and in so doing, preserve decision-making autonomy. Amid this confusion, however, there is a clear drive for countries to adopt standard approaches to risk assessment and regulations that are minimally disruptive to trade. This pressure is reinforced by the actions of GM-exporting countries lobbying weaker governments on a bilateral basis and using the leverage provided by aid and the looming threat of trade sanctions against non-compliant countries. Pressure to fashion a narrow system of biosafety regulation that prioritises market access also comes from the biotech industry itself including of course commodity traders, seeking minimal disruption to the international trade in GMOs.

While more powerful governments may be in a position to accept commitments on their own terms and to resist pressures which they feel go against their national interests, many developing countries are not. They find themselves torn between WTO pressures to open their markets to agricultural imports and resistance from farmers' groups whose livelihoods may suffer from sudden exposure to global markets; they find that their ability to act upon concerns regarding the socio-economic impacts of GMOs on incomes, livelihoods and food security are constrained by international instruments which only deal with the environmental implications of the technology, and they find that global rules on intellectual property rights sit uneasily with indigenous traditions of innovation and ethical concerns regarding the patenting of living organisms, for example. As Zambia, Bolivia and India, among others have also discovered, when a leading GM exporter is also cast in the role of aid donor, the temptation to link the two policy goals is overwhelming, thereby bypassing altogether respect for national choice regarding the technology's acceptability to a society (see Chapter 5).

A critical tension is coming to the fore in the regulation of GMOs, whereby increasing emphasis on public participation in the design of regulations, not least within the Biosafety Protocol itself, sits uneasily with moves by bodies such as the WTO and the OECD which have the effect of reducing scope for government autonomy in responding to diverse public demands. This becomes clear from reading the details of the full evaluation of how governments are interpreting and implementing their commitment to fulfil Article 23 of the Cartagena Protocol regarding involving the public in the design of their National Biosafety Frameworks (Glover et al., 2003; Glover, 2003b). Engaged publics sought to raise issues about why their society needed biotechnology, as well as broader social, ethical, moral and religious issues regarding the technology's development and application which were subsequently found to be 'off-limits' in terms of those issues that were presented to them as legitimate to discuss and which governments were in a position to act upon. Such practices may be constitutive of a democratic deficit played out at the interface of international law responsive to commercial imperatives, aimed at regulating spaces for disruption of the international trade in biotechnology, and national spheres of technology deliberation where potential exists to pose and demand answers to more fundamental questions about the purpose and exercise of social control over technology.

Policies and measures that may be popularly desirable, such as labelling, comprehensive and precautionary forms of risk assessment, forms of trade protection for the poor, restrictions on investment in domestic seed markets or even moratoriums on the trade in GMOs, are increasingly difficult to enforce on the basis that they are incompatible with global trade accords. The extent of this 'disciplining' of domestic autonomy is disputed. Millstone and van Zwanenberg (2003) argue that there is sufficient ambiguity in the respective accords dealing with these issues that developing countries can carve out for themselves a broad domestic priority-driven agenda without fear of direct conflict with WTO strictures. How these conflicts and tensions play out in practice is best understood in relation to specific case studies, which we return to below. Before also considering the democratic implications of how these are resolved or not, it is necessary to understand from whence they derive, to look at why policy autonomy and *regulation of*, rather than *regulation for* business (Newell, 2001), is perceived to be such a threat to corporate interests.

Corporate strategy[1]

Understanding the governance of biotechnology means understanding the reciprocal relationship between intra- and inter-firm decision-making and global decision-making, as multinational firms in particular, are key actors in the global politics of biotechnology. The research behind the development of GM crops, the tests that are undertaken to assess their safety, and the means by which they are distributed, are all strongly affected by the activities of agri-biotech companies. For this reason, questions of corporate strategy and the organisation of the industry tell us a great deal about its *governability* by governments and international organisations, whose responsibility it is to regulate the risks associated with the development and release of GMOs. Policy changes the commercial environment in which investment decisions are taken, and technological innovations and investment practices among leading firms help to construct the policy space available to state managers.

Rather than viewing business interests as homogeneous and monolithic, it is important to look at divisions within capital and the political alliances that firms form as a basis for understanding the ways in which policy choices are framed and decisions taken. There are many important differences among firms with regard to their role in the 'gene revolution' and the public positions they adopt regarding the regulation of biotechnology products. Different corporate strategies engender differences of opinion over intellectual property rights and the protection of plant varieties, as well as issues such as commercial confidentiality and the nature of biosafety regulation.

The corporate strategies of biotechnology firms explain both their market dominance and the stake they have in ensuring a regulatory environment conducive to their commercial interests. Major firms have become integrated horizontally to other sectors (pharmaceutical, chemical) and vertically (from seed production to processing and marketing), some more successfully than others (Chataway, 2001). Mergers and acquisitions consolidate this process, such as the merger between AgrEvo and Rhône-Poulenc to form Aventis CropScience, and are driven by competitive pressures as well as the huge expense associated with biotechnology R&D. There has also been a trend for large 'life science' firms to buy or acquire interests in major seed companies in key markets, exemplified by Monsanto's decision to acquire the seed giant Cargill's international business (Søgaard, 2001). Companies such as AgrEvo made the shift from crop *protection* to crop *production* through biotechnology by acquiring successful biotech firms such as Plant Genetic Systems (PGS) (Bijman, 2001). Concentration in the sector has enhanced both the market share and political significance of these actors. Consolidation of the production chain means increasingly that the same firms have stakes in policy processes governing all parts of the supply chain from research and product development to trade and distribution.

Corporate strategy in the biotechnology sector is continually being redefined and realigned in line with shifting market opportunities and changes of political landscape. This is recognizable in discursive shifts as well as the material changes noted above. Whereas up until a few years ago (some) biotech companies were willing to make inflated claims about the technology's ability to tackle problems of poverty and food insecurity, claims which prompted a backlash from the NGO community and sceptical publics, more recent discourses have emphasised 'farmer's choice'. In India, for example, there is widespread evidence that whichever edicts emerge from the deliberations of regulatory committees in Delhi, the reality 'on the ground' is that farmers are seeking access to GM seeds, largely through black markets. Biotech firms, once seeing this as a threat to their profits, have presented this as evidence that the farmers have already decided that they want the technology, so as to cast decisions made by policy elites which restrict access to it, as anti-democratic. This is an extreme version of *democracy as market choice* and *citizen as consumer* and is not what advocates of democratising biotechnology had in mind in calling for policy more responsive to the needs of poorer groups (Keeley, 2003b).

Unsurprisingly, companies have not be quite so willing to allow their products to be copied, shared and benefited from as these bold claims of 'farmer's choice' or *field-level democracy* might suggest. The high level of start-up capital that is required in crop development drives companies to seek protection for their investments to recover the substantial research and development costs. AstraZeneca claim that their patenting policy 'is aimed at providing each of its new products with an effective shield of valid, enforceable patent and trade mark rights on an essentially global basis to protect it from unauthorised competition during its commercialisation ... [AstraZeneca] will enforce its intellectual property rights fully whenever appropriate. It will also defend vigorously any unwarranted challenge to its intellectual property rights' (AstraZeneca, 1999: 31). For some firms, this defence has taken the form of legal claims brought against individual farmers accused of using the technology without paying for it, the notorious Schmeiser case being perhaps the most high-profile example (Glover and Newell, 2004), but threats to withdraw investment have been repeatedly made to countries such as China, seen to be doing little to prevent illegal use of biotechnology products, and even against such stridently pro-GM nations as Argentina, for failing to protect firm's intellectual property.

The market potential for biotech products, reflected in companies' efforts to exert micro control over their use, and the level of consolidation in the sector, conspire to create powerful actors with strong incentives to ensure that international regulations do not inhibit the trade in GMOs. Pressure to push a product through to commercialisation generates a potential conflict between biotech firms and regulators over commercial confidentiality, however. Applicants have made extensive use of exclusions on grounds of commercial confidentiality and while some information is made public, there remains a tension between transparent regulation and protecting the confidentiality of those being regulated. The reluctance of firms to disclose information about their research and development work on GM crops means that other actors are not in a position to verify the status of claims being made about the stringency of tests being undertaken. Firms have also made clear that they do not trust regulators from the public sector, who as fellow biotechnology researchers and scientists, are regarded as being in competition with the firms. These criticisms have been made repeatedly in the Chinese context in relation to the role of scientists working for Chinese biotech companies such as BioCentury being allowed to oversee approval of other companies' products (Keeley, 2003a; Newell, 2003b).

Global environments

Besides the Cartagena Protocol on Biosafety, an increasing number of developing countries are joining the WTO and have become signatories to a range of agreements on agriculture and intellectual property rights, for example, that impact upon the way they handle biotechnology development and biosafety issues at the national level. Given the potential reach and influence of the agreements, it is unsurprising that the biotechnology industry has mobilised so significantly in order to try and shape them. International legal and policy arenas have become an important site in the battle to secure global market access for biotechnology products.

Regulation can bring order to commercial interactions and lower transaction costs by reducing barriers to trade, creating rules of conduct and preventing the growth of obstacles to commercial development. The Technical Barriers to Trade (TBT) and Sanitary and Phytosanitary Measures (SPS) agreements of the WTO, for example, call for the use of 'sound science' criteria as a basis for evaluating risks that are invoked as a basis for restricting trade. Even the CPB (Article 15) asserts that parties can only withhold consent for the import of a GMO if decisions are based on 'scientifically sound' risk assessment, with the onus of persuasion on the importing country. This narrows the opportunities available to countries to justify restrictions on trade by other means. Reducing the scope for political differences in the way the rules of trade are set is an important aim for the biotechnology industry. They have leant their support to initiatives, such as the guidelines of the OECD (1986, 1992), aimed at developing a harmonised approach to risk assessment that reduces the transaction costs for firms having to meet different standards in different countries. Firms also express a preference for the SPS agreement which calls on members in Article 3 'To harmonize sanitary and phytosanitary measures on as wide a basis as possible' over Codex guidelines or standards on labelling which they describe as 'precautionary, non-science based international food standards' (Mansour and Bennett, 2000: 4).

The pressure to commercialise new products and to get approval for them is also apparent in the debate about the scientific principles that should underpin regulation. For example, the principle of 'substantial equivalence' (SE), developed by WHO, FAO and OECD, is used to compare the risks associated with products containing GMOs with those produced with traditional plant-breeding techniques. According to Phillips and Corkindale (2002: 113), the biotech industry 'has relied on the concept of substantial equivalence in regulatory regimes to justify ignoring the concerns of consumers for most of the GM foods currently in the market'. The OECD has sought to get SE accepted as an international regulatory concept by establishing a programme on the harmonisation of regulatory oversight in biotechnology, reducing scope for competing understandings of risk and assessments of the suitability of a GMO in a range of different agro-ecological environments.

Increasingly, therefore, there are a set of global pressures for establishing common means by which to identify and manage the risks associated with GM products emanating from the OECD, the 'Miami group' and leading industries in the biotech sector. As Levidow et al. (1996: 140) note 'harmonisation efforts gained impetus from many sources: from free-trade imperatives, from applicants operating across national boundaries and ultimately from marketing applications, which stimulated regulators to try to reconcile their data requirements'. For industry it is key that any dispute that may arise are dealt with by an institution sympathetic to their goals. The US Feed Grains Council note, for example,

> . . . The fora in which potential disputes are settled and the agreements they make reference to are key strategic concerns for industry. For consistency, predictability and ultimately lower transaction costs and easier access to markets, industries tend to prefer reliance upon the SPS and TBT agreements of the WTO which describe the least trade discriminating path to risk assessment and standard-setting (quoted in Newell, 2003c: 64).

That industries prefer the WTO as the appropriate venue for settling these issues is perhaps unsurprising given that Barrett and Abergel (2000: 10) note the life science industries are 'hugely influential' in international trade organisations such as the Codex Alimentarius Commission and the Intellectual Property Committee of the WTO. Multinational corporations and trade associations are aware of the role that Codex has been given by the WTO agreements as the venue to resolve disputes over trade in food products. Global trade rules are appealing to industries because they serve to narrow the menu of regulatory choices open to governments. A degree of policy lock-in for governments is, therefore, desirable from an industry point of view. Responding to consumer demands for strict labelling requirements, moratoriums and bans on the import of GMOs all have implications for trade and have raised the spectre of trade disputes, creating tensions between governments' potentially conflicting responsibilities to the public at large on the one hand, and to market actors on the other. The net result is that the autonomy of some governments is reduced and their ability to respond to popular calls restrictions on imports of GMOs is curtailed, as attacks by US trade officials on the GMO regulations of countries such as Sri Lanka, Egypt and Croatia make clear.

Where the stakes are higher, the incentives to employ strong-arm tactics to ensure enforcement are raised. Hence barriers to trade, perceived to be inconsistent with WTO accords, are likely to attract more attention and be dealt with more decisively, than violations of provisions contained in the Cartagena Protocol. Indeed, the overriding aim of many GM-exporting nations and their allies in the business community is to ensure that provisions contained in the protocol which could be invoked to inhibit the free trade in GM products, remain un-enforced or at least under-used. Bilateral pressures to persuade countries not to use them, including the threat of bringing cases before the WTO, are indeed intended to create a general climate of expectation that countries adopting overly restrictive measures from the point of view of GM exporters, can expect repercussions from those whose market share is at stake.

It is unsurprising that biotech companies whose products are the subject of regulation, are heavily involved in the international governance of crop biotechnologies. Systems of governance constructed to address trade concerns nevertheless have important implications for national-level policy choices around biotechnology. The next section explores the negotiation of international obligations and domestic priorities in practice with a particular focus on the extent to which policy choices are bounded by corporate power.

The national politics of development: bounded autonomy

Amid these pressures for conformity, states, some more than others, continue to exercise discretion; a type of *bounded autonomy* with regard to decision-making on biotechnology. This derives from the ambiguity inherent to all legal texts, but magnified in the context of international accords attempting to bridge so many diverse interests, and the inconsistencies that exist between them, the commercial and political importance of some states which provides them with a resilience to 'external pressures', reflecting their potential as export markets or their political power in broader terms, and ties to civil society, which in some settings provide a buffer against the disciplining effects of global institutions and market actors. This autonomy is bounded, nevertheless. It has limits. It is neither infinite nor static.

Many of the examples explored below demonstrate complex patterns of manoeuvring on the part of governments to buy themselves time and space to define for themselves a biotechnology strategy that reflects their priorities, or to advance the commercial prospects of domestic rivals to global biotechnology firms. Advocates of the technology are only too aware of this and the intimate relationship between biotechnology and trade and aid provide ample opportunities for attempting to contain such autonomous endeavours or democratic impulses. As with lobbying at the global level, at the national level, the struggle to secure policy closure around biotechnology is fought via battles about science, ownership and confidentiality, via discourses of choice and of course narratives of competition.

Scientising public debate

With biotechnology we see the essential tenets of modernity and capitalism brought to bear to foreclose broader democratic engagements with what amount, in many settings, to key questions of rights, access and entitlements to food and livelihood security. The scientisation of debate or the 'biotechnologizing' of democracy (Levidow, 1998), has served to restrict debate to a narrow set of scenarios about biotech futures rather than engage broader social and ethical concerns. Even if it is contested outside the regulatory circle, there is little debate among regulators about whether risk is the appropriate frame. Rather, the debate centres on which risks should be the focus of attention with a strong bias towards risks to biodiversity deriving from the Cartagena Protocol. This means that broader ethical questions around the desirability of the technology, or of the power relations it implies, are left off the agenda. Though risk cannot be isolated from ethical and political questions about socially acceptable levels of risk, many regulatory processes are designed in such a way that these broader questions cannot be posed, let alone addressed. Instead, regulations often reinforce a division of responsibility whereby environmental risk assessments are determined by 'objective' science, socio-economic effects are decided by consumer choice and bioethics are provided by professional experts (Black, 1998).

Constructing publics as ignorant, ill-informed and only able to engage with issues of technology on 'emotional' grounds, have been among the devices to create public alienation (Wynne, 2001). This is in spite of the growing popularity of tools for public participation in debates about biotechnology which, according to Levidow (1998: 220), have 'biotechnologized' public participation by narrowing issues to technical problems amenable to neo-liberal risk benefit analysis conducted by specialised experts.

Emphasis upon objectivity and the fact/value distinction has been a recurrent technique for diffusing public disputes over biotechnology regulation. While the European Commission conceded that, in special cases, it may also consider socio-economic aspects of the technology, the European biotech industry has insisted that product regulation should 'assess only safety, quality and efficacy for man and the environment on the basis of objective scientific criteria' (quoted in Levidow and Tait, 1995: 134). 'From industry's standpoint social need should be determined by the free choice of consumers in the market' (ibid).

Secrecy regarding products and the ways in which their safety is being assessed makes the relationship between regulator and applicant discrete and individualised with the latter very much dependent on the integrity of the former. The Environmental Protection Agency (EPA) in the US, for example, accepts the laboratory and field studies of biotech companies, which show no occurrence of harm, as a basis for policy. Voluntary private consultations with the agency before a product is marketed are considered adequate (Levidow, 1999; Hammond and Fuchs, 1999). Participation in decisions about GM crops for most publics takes the form of exercising consumer rights to buy, or refuse to buy, a product that has already been approved for market entry despite the efforts of activists to democratise decision-making through attempting to secure public rights to information, to expose approval processes to public scrutiny.

Exporting regulation through law

If scientising the debate about the scope and nature of regulation serves to exclude publics, state autonomy regarding the essential features of a regulatory framework for biotechnology can also be constrained by the transfer of regulatory systems, tailored to a particular set of commercial needs, to new settings. This can occur through bilateral pressure, which has happened in relation to strengthening IPR protection, for example, through the efforts of international organisations to standardise and harmonise approaches to risk assessment, noted above, or through the advice given to countries about how to design biosafety regulations by 'absentee experts' (Jansen and Roquas, 2005). This has important implications for policy autonomy given that what is imported through policy transfer is not just a tested set of rules and procedures, but a set of values and assumptions about biological processes and prior assessments about which risks are socially acceptable. While there is clearly scope to adapt regulatory models to national needs, it remains the case that certain practices, values and assumptions get internationalised by these means. Capacity-building initiatives hosted by agents such as the Global Environment Facility (GEF) aimed at overseeing the implementation o the Procotol, steer countries towards common approaches to meeting their commitments and in so doing foreclose opportunities for more nationally owned processes of designing *National Biosafety Frameworks* by, for example, reinforcing and reproducing the separation between risk assessment procedures (designed by scientists) and processes of public consultation and participation (led by government around a set of issues that have been pre-determined by experts) (Glover, 2003b).

The example of the contest around intellectual property rights in India yields interesting insights about efforts to reconcile domestic preferences with global obligations. The exemptions India included in its pre-TRIPS legislation in the areas of agriculture and pharmaceuticals in particular, as well as both its large domestic market and competitive potential as an exporter of these products, 'made India one of the targets of US global efforts to tighten IPRs' (Yamin, 2002: 41). When the attempt by the Government of India (GoI) to amend the Patent Act failed in 1995, the US filed and won a case against India brought through the dispute settlement mechanism of the WTO. Various adjustments have been made to the Patent Act to make it TRIPS compliant, including the dilution of provisions excluding plants from patent protection (Ramakrishna, 2003). The government was also obliged to set up a system for the protection of plant varieties given that the 1970 Patent Act excludes plant varieties from protection. Interestingly, however, though India has had to introduce patents for microorganisms and microbiological processes, it has been able to restrict the scope of patents on life by prohibiting patents on cells, cell lines and genes. China's patent law also prohibited patents on plants and animal varieties, even if individual genes can be patented unlike many other developing countries. Amendments were made to China's patent law as part of the accession agreement to the WTO which were introduced in 1992 and 2000. Again, there remain some areas of non or partial compliance concerning, for example, the relationship between IPRs and anti-competitive behaviour.

The extent to which the Cartagena Protocol on Biosafety and the TRIPS accord, in particular, serve to constrain or enable governments' autonomy of action and ability to respond to the needs of the poorest, remains an open question. Clearly the relationship between international rules and patterns of national regulation is not linear. Bureaucracies resist change and regulatory cultures significantly shape the formation of policy in practice through the process of translation. It is often the case that conflicts between national priorities and internationally imposed constraints are, in reality, conflicts between government departments within states over the importance attached to biotechnology development or between governments keen to lure biotech investors and publics sceptical of the technology's benefits. Many governments have been proactive in providing incentives for companies to get involved in research and development of GM crops. Through financial support such as tax incentives for research, soft loans and duty-free imports of equipment, governments are creating markets for biotechnologies by creating the right conditions for GM technologies through policy and infrastructure. Far from a neutral arbiter, governments can use their power either to domesticate the preferences of external institutional and market actors, or use their authority to contest or refract those preferences through the lens of broader understandings of the public interest.

Importing regulation through pressure

Though governments have shown themselves willing to intervene bilaterally on behalf of biotechnology companies, firms themselves play key roles in lobbying for the adoption of particular regulatory models. Particularly interesting here is the way in which business associations serve as transmission belts for the policy preferences of multinational capital. Firms in India that belong to the Confederation of Indian Industry (CII) and the All India Biotech Association, for example, have been strongly supportive of common approaches to risk assessment and the use of principles such as substantial equivalence, reflecting their ties to global industry groupings such as Biotechnology Industry Organization (BIO) (Newell, 2003a; AIBA, 2000). The policy positions adopted by CII, embodied in their White Paper on biotechnology, resonate strongly with the line espoused by BIO with regard to the need for 'sound science', WTO-compatible regulation that is restrictive of the use of socio-economic criteria as a rationale for restricting imports. These arguments have been used to contest the inclusion in India's regulatory system of a requirement that crops are evaluated for their agronomic potential.

A hearing for these positions is secured through privileged channels of institutional access. On the Protection of Plant Varieties and Farmers' Rights bill in India, Monsanto was invited by the Joint Parliamentary Committee to make an oral submission during consultations on the bill in 2000, the only individual company to do so (Seshia, 2002). There is some evidence also of a strong degree of overlap between the Department of Biotechnology (DBT) and the industry whose activities it is meant to regulate, raising concerns about the ability of the same agency to simultaneously be a 'promoter' of a technology and a 'protector' against the risks associated with it. To take just one example, Dr. S.R. Rao Director of the DBT, was formerly involved in developing a strand of biotechnology-related work with Rallis corporation in Bangalore.

The material power that firms wield amid hype about their contribution to the economies of India and China respectively helps to ensure that these channels of institutional access remain open. While access to skilled labour and adequate infrastructure place constraints on where firms can locate, large biotech firms consider themselves to be highly mobile in where they base themselves. This provides them with a degree of leverage over governments anxious to attract investors where they can exercise a powerful threat to move operations elsewhere. Comparisons with China are invoked by bodies such as the CII, as well as individual firms, to underline the fact that if the Indian government does not send out positive signals about biotech development in the country, there are other attractive investment locations that firms could move to.

Regulations have, however, been used to protect domestic industry against foreign competition in certain circumstances. Promoting national capital has been the preferred choice of the Chinese government. The protection of Chinese producers and promotion of China's own biotechnology enterprises has been a key factor in the shift of position from *promotion* to *precaution*. Government programme 863, the platform of China's biotech development, is explicit in its vision to nurture Chinese biotech firms such as BioCentury that, it is hoped, will ultimately be able to compete independently against the likes of Monsanto. This helps to explain the restrictions on foreign investment that other commentators have taken as evidence of a 'cooling' towards the technology per se. A combination of global market imperatives and domestic commercial considerations, what Huang and Wang (2002) refer to as a 'wait and see' strategy, allow China to keep all options about its agricultural development open, preserving autonomy over agricultural futures.

Trade pressure

The high research and development costs and the funding structure of the biotechnology industry which drives companies to lobby for rules minimally disruptive to trade also creates strong incentives for them to push for market access either in their own right or through representations made by government on their behalf. The Chinese market, given its size and strategic importance, has been a particular focus of exporters' attention. A letter to US House of Representatives speaker Larry Combest from the US-China Business Council for example, signed by groups such as BIO, large traders such as Archer Daniels Midland and Cargill as well as individual firms such as Monsanto and Pioneer Hi-Bred International, called on the US government to grant Normal Trade Relations status to China, such as is accorded to other WTO members (The US-China Business Council, 2000). For the US-China Business Council, a key attraction of China's membership of the WTO is that 'The US will have recourse to the WTO dispute settlement mechanisms should China not live up to any of its obligations' (ibid).

The position of biotech exporters is in many ways reinforced by what Breslin refers to as the 'globalising elite' in China, whose agenda, is to lock China into multilateral trade norms and promote domestic political and economic change within China (Breslin, 2003). As an indication of their success in achieving this, the Ministry of Foreign Trade and Economic Cooperation (MOFTEC) announced in May 2002 that more than 2,300 laws and regulations had been amended to comply with WTO rules and 830 abolished since the country joined the trade body on 11 December 2001 (*China Business Review*, 2002). There has also been pressure on the Chinese government from the US, EU and Japan to centralise decision-making on trade issues so that governments and exporters do not have to deal with multiple agencies each with a different mandate. The merging of the State Economic and Trade Commission and MOFTEC under the Ministry of Commerce can be seen as a move in this direction.

Foreign biotech firms have not been left to fight the battle alone by their home countries. Their positions on the various incarnations of China's biosafety regulations have been actively supported and advanced by government officials acting on their behalf. Following China's imposition of a temporary moratorium on GM soybean imports while regulations were developed, President Bush made a high-level visit in February 2002 to persuade the government to keep trading channels open while regulations on biosafety were recast. Chief agricultural negotiator Allen Johnson of the US Trade Representative office said that after 2 days of talks with his Chinese counterparts his mission had been a success and that he expected that China would take necessary steps to 'adjust its regulations so as not to hold up $1 billion worth of annual US soybean exports to China' (Smith and Rugaber, 2002) having received assurances from officials that China would meet its WTO obligations and ensure that US–China trade is not affected by the new regulations (UDS, 2002). Understating the case dramatically, Allen Johnson acknowledged that Bush's visit 'helped' pave the way for the interim settlement.

'Common sense' food security

Beyond these explicit attempts at lobbying through political representation, backed up by material power, the perceived contribution of biotechnology to meeting pressing developmental needs has been viewed as an opportunity to promote the industry as an essential element in the fight against rural poverty in particular.

For biotech companies operating in India, the simultaneous potential for high profitability, global market penetration and the prospect of addressing some of India's food security needs, places them well to argue that their commercial interests coincide with the national interest. The intention of such claims is to fuzzy any distinctions that may exist between notions of what is in the national interest and what is in the interest of leading firms. Hence, while the potential for growth and employment that agricultural biotechnology can deliver is hotly contested in public arenas in India, the prevailing perception among policy elites within government is that biotechnology has great growth potential, reflected in the pronouncements of the Indian Prime Minister and his heads of department in government. Former Prime Minster Atal Bihari Vajpayee stated at the Science Congress in Delhi in 2001 that India's vision included 'shaping biotechnology into a premier precision tool of the future for creation of wealth and ensuring social justice especially for the welfare of the poor' (Herring, 2005). This vision is also shared by the department that is in many ways at the centre of biotechnology regulation in India, the DBT. P.K. Ghosh, former DBT advisor, said that 'we have to push this technology ... it is good for the country' (Newell, 2003a).

Biotech companies have also tried to suggest that the 'crisis' in Chinese agriculture can only be offset by the adoption of their products. David Shi, Monsanto's government and public affairs director, claimed that 'Chinese farmers now earn only 400 yuan per mu (1/15 hectare). But with our Bt technology, they could earn 300 yuan more per mu' (Reuters, 1999). Such claims have been greeted with a greater degree of scepticism by a Chinese government determined to steer the technology's development on their own terms.

The case of China, in particular, suggests that policy autonomy is not co-terminus with democratic debate. Strong states may invoke developmental goals to preserve decision-making space without necessarily doing so in a way which is responsive to public concerns and the needs of poorer groups. Neither China nor India come off well in comparative evaluations of government efforts to enable public consultation and participation in decision-making on biotechnology (Glover et al., 2003). Indeed, often it seems there is an inverse relationship between autonomy achieved by virtue of political power and commercial importance and levels of democratic public engagement. As Glover (2003b: 24) concludes: 'in cases where there are few vested interests or intensely opposed interests ... governments may have greater freedom to open the doors to a more inclusive deliberative process. Sadly, therefore ... in practice public participation is much more likely when the stakes are lower – in other words, when it matters less'.

Conclusion: democratising biotechnology?

This chapter has shown how the degree of policy autonomy available to state decision-makers regarding biotechnology is circumscribed by international rules responsive to the needs of GM exporters. Increasingly integrated, and with global control over the supply chains that they administer, biotech multinationals are at the heart of bargains with policy-makers over GM regulations. Pressures for a short-term return on investment, concerns over commercial confidentiality and the organised efforts of industry groups to ensure the adoption of harmonised and universalised approaches to risk assessment that are minimally disruptive to trade, conspire to constrict the scope for meaningful public engagement with decisions about how to assess and manage the risks associated with GMOs.

The notion of 'bounded autonomy' helps to convey the patterns of accommodation and resistance that characterise the interface between domestic agenda-setting and global pressures for conformity, be they political or market-driven. Where those boundaries are drawn appears to be a function of state power, market positionality, and the role of civil society. These help determine vulnerability to pressure from powerful governments, dependence on overseas markets and the extent to which supportive coalitions exist in civil society to buffer government positions in the face of international criticism. As the examples above make clear, countries face very different dilemmas as a result of this diversity.

The account developed here is at odds with those readings of global biotech politics which suggest state impotence in the face of biotechnology companies and global institutions such as the WTO. Instead, the patterns of manoeuvring and the politics of translating commitments embodied in international texts in ways which reflect domestic needs have been highlighted. It is also at odds with those accounts which blame the 'slow' pace of take up of biotechnology in developing countries on the overwhelming pressure of anti-GM activists within developing countries but acting as the mouthpiece for elite Western-based environmental interests (Paarlberg, 2001). While civil society activism has been an important part of the story in India, it has not been crucial in China, which has also latterly adopted a more cautious position regarding commercialisations. Moreover, in spite of strong civil society resistance, India has moved ahead with commercialisations suggesting that anti-GM activists do not wield the sort of power they are assumed to. Finally, accounts which suggest international legal texts provide ample scope for wide interpretation and, therefore, discretion in policy responses (Millstone and van Zwanenberg, 2003), while correct, overlook the extent to which pressures, political and commercial, are brought to bear upon countries, particularly developing countries, to interpret and act upon those commitments in ways which conform to the needs of GM exporters rather than domestic constituencies. As these authors acknowledge: 'How much discretion governments can exercise is not something that can be determined solely by scrutinising the texts and rules of the WTO and the CBSP' (ibid: 660).

Taking the example of the ability of industries to shape biotech regulations in developing countries, I have shown how the form and scope of 'bounded autonomy' is a function of a number of key variables. One is a government's own perceived interest in the issue and the extent to which biotechnology is prioritised as an area of strategic economic importance. For a country like Argentina, seeking to become a global leader in GM products, the assistance of industries in designing regulations to support this goal is, of course, welcome. Other countries are more ambivalent about their relationship with biotech firms because they are unsure of where to position themselves in the global marketplace for agricultural products. Where a country is located within the supply chain also has a bearing on their ability to resist pressures to alter regulations according to the preferences of buyers. Developing country firms seeking to export to Europe or North America, may be forced to meet the regulatory requirements of those countries. To gain a share of global markets, countries such as China and India have to apply measures in consistent and non-discriminatory ways in terms of the demands they make of foreign as opposed to domestic capital to accord with global trade rules.

Regulations are not only exported through bilateral governmental pressure, or through efforts at harmonisation orchestrated by international organisations, therefore, but also through the vehicle of the supply chain and inter- and intra-firm trade. It is clear from this other work (Newell and Mackenzie, 2003) that perhaps the strongest and most effective pressures towards policy conformity in the global politics of GMO regulation derive from informal, bilateral, trade, aid and market-based pressures. These have been used to constrain and discipline those countries adopting regulatory models threatening to the interests of biotech exporters and seem to act as a far more immediate catalyst to action than the well-intended but abstract commitments contained in the texts of global legal instruments.

Note

1. Parts of the section draw on Newell (2003a).

5

The Political Economy of Food Aid in an Era of Agricultural Biotechnology

Jennifer Clapp[1]

In 2002 the United States sent significant quantities of food aid, in the form of whole kernel maize, to Southern Africa in response to the looming famine in the drought-stricken region. It soon became apparent that the aid contained genetically modified organisms (GMOs), though the recipients had not been notified prior to the shipments being sent. Many Southern African countries initially refused to accept the genetically modified (GM) food aid, partly as a health precaution, and partly on the grounds that it could contaminate their own crops, thus hurting potential future exports to Europe. A number of the countries eventually accepted the food aid provided it was milled first, but Zambia continued to refuse even the milled maize. The United States argued that it could not supply non-GM food aid, and it refused to pay for the milling. The United States then blamed Europe's moratorium on imports of GM foods and seed for contributing to hunger in Southern Africa.

This incident highlights a new aspect of the recent global predicament over how the international movement of GMOs should be governed. Although there has been much analysis of this question with respect to commercial trade in recent years, particularly regarding the adoption of the Cartegena Protocol on Biosafety (for example, Falkner, 2000; Newell and Mackenzie, 2000; Bail, Falkner and Marquard, 2002), the literature on food aid has not kept up with these new developments. Recent academic analyses of food aid have paid little attention to the question of GMOs. The literature on food aid has focused mainly on the motivations for donating food aid, and its potential as a development tool. Some have argued that although economic and political considerations are present to some degree in the motivation for giving food, today it is mainly given as part of a development regime that aims primarily to promote food security and rural development rather than as a means to serve the domestic economic and political interests of donor countries (Uvin, 1992; Hopkins, 1992, 1993; Clay and Stokke, 2000).

In light of the changes in global agriculture over the past decade, especially the rise of the United States as a major producer of GM crops, it is
important to re-examine the political economy of food aid. There appear to
be strong economic motivations for the United States to pursue the food aid
policy described above, as well as scientific motivations, not addressed by
the earlier food aid literature. Both of these motivations are highly politi-
cized. Europe has not followed the lead of the United States on GM food aid
policy. The divergence of the policies of the EU and of the United States on
this issue may well lead to interesting politics in the coming years in the
international battle over GMOs. This time, however, the debate looks set to
be played out globally with some of the world's poorest countries as unwit-
ting participants in the conflict.

Why revisit food aid politics?

The modern era of food aid was instituted in the United States in 1954 with
the passage of U.S. Public Law (PL) 480. Since then food aid has been an
important feature of U.S. assistance to developing countries, though its role
has changed over time. In the 1950s, food aid accounted for nearly a third
of U.S. agricultural exports, whereas in the mid-1990s it was closer to 6
percent (Christensen, 2000: 256). Food aid under PL 480 is given under
three different categories of assistance. Title I is government-to-government
aid in the form of concessional sales with the express aim of opening new
markets for U.S. grain. Title II is grant food aid distributed in emergencies.
Such aid can be distributed via nongovernmental organizations (NGOs) and
the World Food Programme (WFP). Lastly, Title III is government-to-government grants of food aid for development activities (USAID, 1995).

From its origins, U.S. food aid was largely seen as a multipurpose tool. On
the surface, the idea of the PL 480 was to provide the world's food-deficit
countries with food from the United States as part of a broader humanitarian effort. It was also clearly a mechanism for surplus disposal and export
promotion in the United States. It created a market for surplus food and as
such it had the effect of raising U.S. domestic prices for grain. Further, it was
hoped that shipping free or concessionally priced U.S. grain to poor countries would create new markets in the future for commercially traded U.S.
grain (Friedmann, 1982; Ruttan, 1993). U.S. food aid was, however, soon
seen as a political tool as well. The United States had even gone so far as to
amend PL 480 in the 1960s to explicitly tie the donation of food aid to political goals, in particular to favor noncommunist countries (Wallerstein,
1980). Other countries followed the United States in giving food aid, including Canada, which began its program in the 1950s, and the European
Economic Community, which began to give food aid in the late 1960s (for
example, Cathie, 1997; Charlton, 1992). Canada and the European donors
have been less overtly political and economic in their rationale for food aid,

although some surplus disposal mechanism has been part of their food aid donations at various times.

The UN World Food Programme (WFP) was set up in 1963 as a multilateral channel for food aid from donor countries. In its early years the WFP distributed around 10 percent of food aid, whereas today nearly 50 percent of all food aid is channeled through the WFP (IGC, 2004). The Food Aid Convention (FAC) was first established in 1968 as part of the Kennedy Round of GATT negotiations, and was attached to the International Wheat Trade Convention (International Wheat Council, 1991). The major donor countries are FAC members, and the agreement sets out minimum amounts of food aid per year each donor is to give (denominated in tonnes of wheat). Re-negotiated periodically, the FAC now stipulates that food aid can be given either in kind or in cash equivalent, and that other commodities apart from wheat can also be given (but they are still measured in wheat equivalents). Today the donor members of the FAC include Argentina, Australia, Canada, the European Community and its member states, Japan, Norway, Switzerland, and the United States (Food Aid Convention, 1999). From the early days of food aid to the present, the United States has remained the principal donor country, and gives its aid primarily in kind. Other donors have over the past decade increasingly given their food aid in the form of cash, which is directed toward third-country purchases or local purchases in the recipient region.

Since the 1980s food aid policies in the United States have been reformed significantly, with the overt political goals removed from the PL 480. And in the mid-1990s, the surplus of grain in the United States diminished, making the surplus disposal element of food aid appear to be less significant than it was in the past (Christensen, 2000). In the European Union, food aid policies since the 1990s have focused on giving aid in the form of cash to finance food distribution programs for food obtained through local or third-party sources, rather than in kind. The EU regulation of requiring the shift toward cash-based food aid in 1996 reinforced this policy (Barrett and Maxwell, 2005; European Union, 1996), which for the EU was largely in response to studies that showed that cash spent on local purchases of food in aid-recipient regions boosts the local economies and allows for much more flexibility in terms of sourcing culturally appropriate foods (Clay and Stokke, 1991).

These policy shifts prompted some to argue that food aid by the late 1980s and early 1990s had in fact become largely a development tool, with the motivations for donating food governed more by the existence of an international regime (and desire to cooperate) than by donor economic and political considerations, which is in line with a liberal institutionalist perspective on international relations (Uvin, 1992; Hopkins, 1992, 1993). In other words, a 'depoliticized' food aid regime was seen to have emerged, which was not merely serving the interests of the donors but rather was promoting international development. This was especially the case for European donors. Uvin (1992: 307–8) argued that by the early 1990s most

EU food aid and about 60 percent of U.S. food aid was clearly not driven by economic or political motivations. Other, more recent studies have made similar argument . Neumayer (2005), for example, argues that in the 1990s although U.S. economic and military strategic interests had some influence on food aid donations for longer-term food aid, it had an insignificant influence on emergency relief.

In view of the development of agricultural biotechnology in recent years, I argue that it is important to re-examine the political economy of food aid. Important factors influencing donor motivations in an age of agricultural biotechnology are not adequately considered in the food aid literature. Economic factors may once again be key motivating factors for food aid policy. These factors are especially important to consider given the growing corporate concentration in the agricultural biotechnology sector and their close ties to U.S. government agencies, as well as the U.S. dispute with the EU at the WTO over its 998–2004 moratorium on GMOs. Moreover, new factors that may influence food aid policy must also be considered, such as the scientific debates over the safety of GM food. These factors appear to be influencing food aid policies on the part of both donors and recipients, and they are highly political.

GMOs in food aid

GMOs have been present in food aid since GM soy and maize were initially approved for production in the United States in the mid-1990s. Their presence in food aid was inevitable for a number of reasons. U.S. food aid is predominantly given in kind and is made up of food (mainly wheat, corn, and soy) grown in the United States, which is by far the largest producer of GM crops, accounting for over 50 percent of the global acreage planted with GM varieties. Between 1996 and 2004 the global area planted with GM crops increased 47-fold to cover 81 million hectares in 2004 (James, 2004: 4). In 2004, 85 percent of soy and 45 percent of maize grown in the United States were GM varieties (Pew Initiative on Food and Biotechnology, 2004). Moreover, there is not a segregated system for GM and non-GM crops in the United States, which has led to widespread cross contamination. This is important, as the United States accounts for 60 percent of all food aid donations.

Although negotiations began in 1996 on a protocol on biosafety under the Convention on Biodiversity to address the safety issues related to trade in GMOs, there was little attention paid to their presence in food aid transactions at that time. It is not surprising, then, that when some food aid donations were discovered to contain GMOs in 2000, many were caught unaware. Both USAID and the WFP had sent shipments of food aid containing GMOs, amounting to some 3.5 million tonnes per year (Lean, 2000; Pearce, 2003:). Such shipments were often in contravention of the national regulations of the recipient country.

Ecuador was the first-known developing country to receive food aid containing GMOs in a shipment of soy from the United States and channeled through the WFP. The product was eventually destroyed following complaints by Ecuador (FOEI, 2003: 5). There were also cases of GMOs being sent in food aid shipments to Sudan and India in 2000. In 2001, GMO soy was found in food aid shipments sent to Columbia and Uganda. Food aid maize from the United States that contained GMOs was also reportedly sent to Bolivia in 2002, despite that country's moratorium on the import of GMO crops. The GMOs found in the Bolivian aid contained StarLink corn, which is a modified form of corn that was not approved in the United States for human consumption (it was approved as animal feed), but which nonetheless managed to enter the human food system in the United States in the fall of 2000. NGOs claim that despite the fact that when StarLink was found in the U.S. food supply it was immediately removed from the market, the United States did little to remove the maize from Bolivia. In 2002, WFP also sent GM corn seed as food aid to Nicaragua and Guatemala. This was particularly controversial in Nicaragua, a country that is a centre of origin for corn (FOEI, 2003: 6–7; ACDI/VOCA, 2003: 6–10).

By mid-2002, there were enough incidents of GMO food aid to have made the donors fully aware of the issue. Recipient countries expressed concern about the potential health and environmental impacts of GMOs, including allergenicity, and outcrossing of GMOs with wild relatives (if the grains are planted rather than eaten) that could reduce biodiversity by contaminating and driving out local varieties. Once GMOs are released into an environment, they are difficult, if not impossible, to remove. Food aid, when given in whole grain form, is often planted by local farmers, who may have exhausted their seed supply as food in times of crisis (GRAIN, 2002). The fact that GM crops have not been approved in many countries, including in the EU, which had placed a moratorium on their imports and new approvals of GM crops in 1998, fueled many of these concerns, especially for those countries that have export markets in the EU.

Until mid-2002, the food aid shipments identified as containing GMOs were mainly to areas, which, although in food deficit, were not facing an acute food shortage. This changed in mid-2002 with the looming famine in Southern Africa. Some 14 million people in six countries faced imminent food shortages and potential famine at that time. A number of factors were seen to have precipitated this situation. Drought and floods were identified as one of the immediate causes. However, underlying factors were just as important. These include the high prevalence of HIV/AIDS in the region, conflict in Angola (and refugees from Angola in neighboring countries), domestic agricultural policies, and the impact of trade liberalization under structural adjustment in some countries in the region (Oxfam International, 2002: 6). It was the worst food shortage faced by the region in 50 years.

In response to this crisis the United States sent 500,000 tonnes of maize in whole kernel form to the region in the summer and fall of 2002 as food aid. The WFP estimated that around 75 percent of food aid to the region at that time contained GMOs (WFP, 2002: 4–5). The countries that received the shipments were Zambia, Zimbabwe, Malawi, Swaziland, Mozambique, and Lesotho. The aid was channeled through the WFP and NGOs. When the countries discovered that the aid contained GMOs, they were forced to consider whether they wanted to accept the aid. Zimbabwe and Zambia said they would not accept the food aid at all, whereas Mozambique, Swaziland, and Lesotho said they would accept it if it was milled first. Malawi accepted it with strict monitoring to ensure that its farmers did not plant it. Zimbabwe eventually said it would accept, it if it was milled and labeled first (Institute for the Study of International Migration, 2004: 16–7).

Zambia stood firm in not accepting it for its own people. It did eventually accept it in milled form but only for the 130,000 Angolan refugees in camps within its borders, and not for the general population (Bennett, 2003: 29). Zambia expressed its concern that any health problems that might arise from eating GMOs would be too costly for the country to address. Since the Zambian diet consists of far more maize than the diets of North American consumers, such health problems may not be foreseen. Moreover, Zambia does have some maize exports to Europe, and contamination of its maize with GMOs could affect its exports. The value of Zambian exports of maize in 2002, for example, was US$2.23 million, according to the FAO (FAO, 2004b). The WFP scrambled to find non-GMO aid for Zambia, which had some three million people at risk of starvation. In the end, the WFP was only able to source about one half of the necessary 21,000 tonnes of maize needed for Zambia (Crilly, 2002: 12).

The WFP responded by saying that it was impossible to mobilize non-GMO food aid fast enough. The WFP made it clear that it respected the right of the countries to refuse to accept the aid, and it did what it could to organize the milling of the maize for those countries that would accept it in that form, and to source non-GMO aid for Zambia. The WFP had to quickly arrange local milling, and in the case of Zambia, it had to remove shipments that had already been delivered. The milling did provide the WFP with the ability to fortify the grain to raise its micronutrient content, however, which was seen as an unexpected benefit. Further, the WFP did manage to solicit donations from nontraditional donors of food aid, including a number of developing countries (Bennett, 2003: 29).

The response of the United States was much more defensive. The United States refused to mill the maize before sending it to the region, claiming that it was too expensive to do so – raising the cost of the food by as much as 25 percent – and that it would reduce its shelf life. It also initially refused to send non-GM varieties of corn to the region, claiming that it was impossible to source non-GM crops from the United States. It refused to give cash instead of in-kind aid, on the grounds that it has traditionally given in-kind aid and has done so for 50 years. The United States did, however, stress that it would respect the wishes of the countries that did not want GMO food aid sent to them. The U.S. position was clearly indicated in a USAID website 'questions and answers' on the GMO food aid crisis (USAID, 2003c). The United States did eventually give Zambia a donation of GM-free maize of some 30,000 tonnes, however, after heavy international pressure to do so (Mellen, 2003).

The United States also blamed Europe for the crisis, saying that its moratorium on approval of seeds and foods containing GMOs was stalling efforts to promote food security (Grassley, 2003). In the midst of the African crisis the United States began to seriously consider launching a formal complaint at the World Trade Organization (WTO) against the EU moratorium on GMOs, claiming that it was in contravention of WTO rules. Although the WTO rules do allow countries to ban imports of a product based on food safety concerns while the country seeks further scientific evidence, the United States argued that 5 years was sufficient time and that no such evidence had been gathered. The concern of the United States was that the EU position was influencing too many countries, including those in Africa (Borlaug, 2003).

Throughout the fall of 2002 and early in 2003, the United States put pressure on Europe to remove its moratorium. The United States finally launched a formal complaint against the EU at the WTO in May 2003. Argentina and Canada joined the formal challenge (Brack, Falkner and Goll, 2003). At the time U.S. President George Bush stated: 'European governments should join – not hinder – the great cause of ending hunger in Africa' (quoted in Denny and Elliott, 2003). Egypt, which has an active agricultural biotechnology research program, was initially listed as a co-complainant but withdrew because Europe is a very important market for its exports of fresh fruits and vegetables. The United States had hoped that having Egypt on board would strengthen its argument that GM crops are beneficial to Africa. The United States retaliated against Egypt for its withdrawal by pulling out of talks on a free trade agreement with it (Alden, 2003).

When the United States launched the dispute, the EU issued a press release stating its regret over the U.S. decision to take action on this case. It criticized the United States for using the refusal of African countries to accept GM food aid to pressure the EU:

> The European Commission finds it unacceptable that such legitimate concerns are used by the U.S. against the EU policy on GMOs . . . Food aid to starving populations should be about meeting the urgent humanitarian needs of those who are in need. It should not be about trying to advance the case for GM food abroad . . . (European Union, 2003)

Amid the Southern African crisis, the European Commission specifically requested the WFP to purchase only non-GM maize as food aid with the money the EU donates for such assistance (European Union, 2003).

In mid-2003, another dispute over GM food aid emerged. The United States had pressured Sudan to accept GM food aid, despite its recently passed legislation that requires food aid to be certified GM-free. In response to U.S. pressure, Sudan issued a temporary 6-month waiver to this legislation in order to give the United States more time to source GM-free food aid. In March 2004, however, the United States threatened to cut Sudan off from food aid completely (African Centre for Biosafety et al., 2004). This prompted Sudan to extend the waiver to early 2005.

Unclear rules on international trade in GMOs

How is it that such massive shipments of GM food aid could have been sent without the recipients knowing about it before it was sent? The rules regarding trade in GMOs were unclear at both the global and local levels between the mid-1990s and 2003. This was the very period when much of the controversy over GMO food aid was at its highest.

As of mid-2002 when the Southern African crisis erupted, only a few developing countries had any domestic legislation dealing with GMO imports, let alone GMO food aid. At that time, the only sub-Saharan African countries with biosafety laws in place were South Africa and Zimbabwe, although since then a number of other countries have begun to develop policies on GMO imports. Zimbabwe had a biosafety board to advise on GMOs, but it has not approved any GM crops for commercial release. South Africa is the only sub-Saharan African country to approve the commercial planting of GM crops. In July 2001, the Organization of African Unity (OAU, now the African Union) endorsed a Model Law on Safety in Biotechnology that takes a precautionary approach to biotechnology and calls for clear labelling and identification of GMO imports (Baumüller, 2003). This model legislation was designed to guide countries in formulating their own national laws on biosafety and to provide a way to develop an Africa-wide system for biosafety. A number of African countries have now adopted some form of biosafety legislation,[2] though the model legislation has not always been a strong influence on these new laws (see African Centre for Biosafety, 2005).

In response to the Southern African crisis in 2002, the Southern African Development Community (SADC) established an advisory committee to set out guidelines for policy on GMOs in the region. These guidelines stipulate that 'food aid that contains or may contain GMOs has to be delivered with the prior informed consent of the recipient country and that shipments must be labeled' (FOEI, 2003: 9). But such guidelines were not available at the time of the crisis. Other regional responses include efforts by the Common Market for Eastern and Southern Africa (COMESA) to develop a regional policy on GMOs, including food aid. And the New Partnership for Africa's Development (NEPAD) decided in mid-2003 to establish a panel to advise African countries on biosafety and biotechnology as a means to try to harmonize regulations on these issues across Africa (Baumüller, 2003: 14).

At the international level, rules on biosafety and trade in GMOs were also not clear prior to 2003. The Cartegena Protocol on Biosafety, which governs trade in GMOs, was negotiated between 1996 and 2000 (when it was adopted), but was not in legal force until September 2003. The rules of the Protocol state that GMOs[3] intended for release into the environment (seeds) in the importing country are subject to a formal advance informed agreement (AIA) procedure for the first international transboundary movement to a country. Importing countries can reject these shipments based on risk assessment. Genetically modified commodities (living modified organisms intended for food, feed, or processing) are exempted from the formal AIA procedure, and instead are subject to a separate form of notification in the form of a Biosafety Clearing-House, which is an internet-based database where exporters are required to note whether shipments of such commodities 'may contain' GMOs. Importers can also reject such shipments based on risk assessment. In both cases, parties are given the right to make decisions on imports on the basis of precaution in cases where full scientific certainty is lacking (text of the Cartagena Protocol on Biosafety). These rules did not cover food aid donations shipped prior to the Protocol's entry into force. And now that it is in force, these rules apply only to those countries that have signed and ratified the agreement. The United States and Canada, two of the major food aid donors that grow significant quantities of GMOs, have not yet ratified the Protocol, and thus are not bound by its rules, unless required to comply with import rules established by members of the Cartegena Protocol.

The Codex Alimentarius Commission, which sets voluntary international guidelines on food standards, was from the late 1990s trying to address questions of biotechnology and food safety. In 1999, the Codex established a special Task Force on Biotechnology to address the wider concerns expressed about biotechnology and food safety, especially those related to risk analysis. The Task Force only released its guidelines in mid-2003. Although they are voluntary, the standards are considered a benchmark for international trade under the WTO. The biotechnology guidelines adopted include safety evaluations prior to marketing of GM products, and measures to ensure traceability in case a GM product needs to be recalled (Codex, 2003). But because these guidelines were not in place at the time of the food crisis in Southern Africa, nothing was done to ensure these guidelines were followed for food aid.

The WFP did not set an explicit policy on how to deal with GMOs until mid-2003. Its policy has long been to give food aid to countries in food deficit if the food met requirements for food safety by both the donor and the recipient. But if neither the recipient nor the donor had a policy of noti-fication, it was difficult for the WFP to track them. It defended its lack of a GMO policy prior to that date by stating that 'none of the international bod-ies charged with dealing with foods derived from biotechnology had ever requested that the Programme handle GM/biotech commodities in any special manner for either health or environmental reasons' (WFP, 2002: 5). Because of media attention to the issue, and claims that the WFP was negli-gent, the WFP decided to establish a formal policy to deal with GMOs in food aid in 2002, which was finalized in 2003. The new policy asks recipi-ent country offices of the WFP to be aware of, and comply with national reg-ulations regarding GM food imports. It also maintains its original policy that it will only provide food as aid that is approved as safe in both donor and recipient nations. Countries that clearly state that they do not wish to receive GM food aid will have their wishes respected. The WFP stated that it will still accept GM food aid from donors, but will also respect the wishes of donors who give cash in lieu of in-kind aid and request that the money not be spent on GM food (WFP, 2003).

Unpacking motivations for GM food aid policy

What explains the widely divergent positions of donors on GM food aid, specifically the United States and the EU, and the rejection by recipient coun-tries? As mentioned above, much of the food aid literature views the current donor motivation to give food aid as not being driven as much by econom-ic and political goals as had been the case in the past, especially with respect to emergency aid. In this section, I argue that we need to revisit this issue. In an age of agricultural biotechnology, new issues must be considered as hav-ing an influence on food aid policy, primarily the scientific debate over the safety of GM food. Further, economic and political incentives, inextricably tied to corporate interests in agricultural biotechnology, appear once again to be important factors behind the U.S. position on GM food aid in particular.

Debates over the science of GMOs: differing interpretations of risk and precaution

The Southern African crisis fueled an already existing scientific debate over the safety of GMOs and their role in promoting food security. The debate has largely been over whether there is sufficient risk associated with the planting and consumption of GM crops and foods to warrant precaution with respect to their adoption. In the media accounts of the GM food aid incidents in Southern Africa, this scientific dimension has tended to dominate the expla-nations for the policies pursued by the donors and the recipients.

The North American position on the safety of GM foods and crops is that there is minimal risk attached to them, and that because of this a precau-tionary approach in their adoption is not warranted. In both the United

States and in Canada, regulatory procedures for GMOs are premised on the logic that if the developer of a GM crop or food can demonstrate that it is 'substantially equivalent' to a conventional counterpart, the GM crop or food does not require an extensive risk assessment prior to its approval (Prakash and Kollman, 2003: 625). Ongoing scientific uncertainty with respect to the risks of GMOs does not automatically invoke a precautionary approach in these countries. In other words, agricultural biotechnology products are assumed to be innocent until proven guilty. It is further argued that the benefits of GM crops, in terms of higher yields and easier management of weeds, far outweigh the (known) risks associated with them (Paarlberg, 2000: 24–38). The United States and Canada view their approach to the regulation of agricultural biotechnology as being firmly grounded in 'sound science.'

The approach to regulating agricultural biotechnology products is very different in the EU and in many developing countries. The EU's interpretation of the potential risks associated with GMOs is much more precautionary. It views the existence of scientific uncertainty with respect to the safety of agricultural biotechnology as sufficient reason to take more time to evaluate the full range of potential risks associated with these products prior to their approval. Before such products can be approved, they must be subject to a rigorous scientific risk assessment (Isaac and Kerr, 2003b: 1086–90; Prakash and Kollman, 2003: 626). In this sense, agricultural biotechnology products are assumed to be guilty until proven innocent. Many developing countries lack a regulatory structure to approve agricultural biotechnology products, and for this reason they have tended to favor the EU approach that applies precaution in the face of scientific uncertainty. Further, there is widespread sentiment in Europe and in many developing countries that the potential risks, such as the potential for outcrossing with wild relatives and creating new allergens, do not outweigh the possible benefits, which reinforces the precautionary mood in those countries.

The different interpretations of the science and risks of GMOs helps to explain the widely divergent positions to GM food aid among the United States, the EU, and the recipient countries. The hostility on the part of the United States toward those countries that rejected the GM food aid, and the placing of blame on Europe, are partly products of these different viewpoints. In particular, the United States considers the EU regulatory system, which is much more precautionary, as being too 'emotional' and not scientifically based. The United States would much rather see its own regulatory style, rather than the EU approach, adopted in developing countries that currently lack a regulatory framework. This U.S. attitude is evident in the comments made by U.S. Senator Chuck Grassley, at a speech to the Congressional Leadership Institute in March 2003, just prior to the launch of the trade dispute against the EU:

By refusing to adopt scientifically based laws regarding biotechnology, the EU has fed the myth that biotech crops are somehow dangerous . . . The European Union's lack of science based biotech laws is unacceptable, and is threatening the health of millions of Africans (Grassley, 2003).

The refusal of the GM food aid on the part of the Southern African countries can also be viewed as a reflection of their position in the scientific debate, as many of the comments made by African leaders when rejecting the aid made this specific link. For example, Zambian President Levy Mwanawasa expressed his concern that GM food aid was 'poison,' stating 'If it is safe, then we will give it to our people. But if it is not, then we would rather starve than get something toxic' (quoted in Dynes, 2002: 12). The Zambian government did authorize a scientific delegation to study the issue, which was sponsored by the U.S. government and several European countries. This delegation traveled to South Africa, a number of European countries, and the United States. The eventual report from the delegation, which came in the fall of 2002, cautioned against the acceptance of GMOs in Zambia, much to the disappointment of the United States (Carroll, 2002; Institute for the Study of International Migration, 2004: 20).

Economic motivations

Although the different interpretations of risk and precaution are clearly relevant in explaining motivations for GM food aid policy, they are not the only significant factors. In an age of agricultural biotechnology, it appears that economic considerations are re-emerging as explanatory factors for food aid policy, at least on the part of the United States.

Throughout the history of food aid, surplus disposal has remained important for the United States (Diven, 2001: 471). Because stocks have declined over the past 50 years, particularly in the past decade, however, some say that surplus disposal is no longer as important as it once was (Christensen, 2000: 257). But the advent of GM food aid may be reviving and reinforcing the surplus disposal aspect of U.S. food aid. The European moratorium on GMO imports has meant a significant loss of markets for U.S. grain. For example, since 1998 the United States has lost around US$300 million *per year* in sales of maize to Europe (Brack, Falkner and Goll, 2003: 3). Some 35 countries, comprising half of the world's population, have rejected GM technology, which also closes the market opportunities for GMO-producing countries to export their products. In addition to the EU, Australia, Japan, China, Indonesia, and Saudi Arabia, also refuse to approve most agricultural biotechnology for domestic use and import (Dauenhauer, 2003). Because of the loss of these markets, the United States may well be looking for other outlets for its GM maize.

The inability to find export markets for its GM grain may be a principal reason why the United States continues to insist on giving its food aid in kind, rather than in the form of cash. Both the FAC and the WFP encourage food aid donations in cash rather than in kind, and the EU has been pushing to have cash-only donations of food aid incorporated into WTO rules. In the case of Southern Africa, the United States was the only donor that gave food aid in kind rather than as cash (WFP Official Richard Lee quoted in Greenpeace UK, 2002). This may be in part due to the preferences of the strong grain lobby in the United States. In a letter to the U.S. Trade Representative on this issue, the National Wheat Growers Association stated: 'We wish to assure you that producers across the nation are strong supporters of humanitarian programs, but will not be willing to support cash-only programs' (National Association of Wheat Growers, 2004).

A second potential economic motivation for the United States in giving GM food aid, not unrelated to surplus disposal, is to subsidize the production and sale of GM crops, and the agricultural biotechnology sector more broadly, which is dominated by U.S. transnational corporations (TNCs). Some 80 percent of funds for the PL 480 program are spent in the United States (Oxfam America, 2003: 8). In 2000, it was reported that Archer Daniels Midland and Cargill, two of the largest grain trading corporations, were granted a third of all food aid contracts in the United States in 1999, worth some US$140 million (Walsh, 2000: 18). The U.S. Department of Agriculture (USDA), which is responsible for regulating biotechnology in the United States and which also oversees the Title I food aid, works in close cooperation with the agricultural biotechnology industry (Stapp, 2003). One example of this cooperation is the 2002 U.S. Farm Bill that provided funding for the USDA to set up a biotechnology and agricultural trade program with the aim 'to remove, resolve or mitigate significant regulatory nontariff barriers to the export of United States agricultural commodities' (Section 1543A of the Farm Bill, cited in ACDI/VOCA, 2003: 5). USAID, which is responsible for Title II and Title III food aid, also actively promotes the adoption of agricultural biotechnology in the developing world through educational programs, giving some US$100 million for that purpose in recent years (*The Ecologist*, 2003: 46). This includes USAID funding for private–public partnerships, such as the African Agricultural Technology Foundation (AATF, 2002) and the Agricultural Biotechnology Support Project (ABSPII, 2004) both of which have heavy participation from TNCs in the agricultural biotechnology industry. These initiatives seek to promote the use of agricultural biotechnology in the developing world through research, education, and training, and they also acknowledge that they hope such efforts will open new markets in the future (Mellen, 2003).

Critics see such efforts as a means by which the United States is trying to pave the way for the introduction of pro-GM legislation to facilitate the export of GM crops and seeds around the world (Kuyek, 2002; Mellen, 2003). For many the position of the United States in the Southern African crisis, especially its refusal to mill the GM grain and its attack on the EU regulatory structure, was seen as a deliberate strategy to spread GMOs as far and as wide as possible, in order to break the remaining resistance to the technology (Glover, 2003a; Kneen, 1999).

Economic considerations in the EU must also be taken into account in unpacking donor motivations. The position of the EU on GM food aid is very much tied to the WTO trade dispute over GMOs more broadly. The United States had been pressuring the EU to lift its moratorium on approvals of GM crops and foods before the crisis hit in Southern Africa, and, therefore, it is not surprising that the EU position was in opposition to that of the United States. Tied up in this broader dispute is the question of export markets for the EU as well. It may be that the EU is seeking to solidify trade relations with developing countries by creating a non-GM market that would exclude the United States. The EU has also been pushing for several years now for cash-only food aid to be written into WTO rules as part of the ongoing talks on the revision of the WTO's Agreement on Agriculture. The EU sees the United States in-kind food aid, and Title I sales of food aid, as unfair subsidies to the U.S. agricultural industry, and wishes to see these removed in exchange for its own subsidy reductions. This helps to explain the EU's criticism of in-kind food aid (Clapp, 2004).

On the recipient side, economic considerations are also important in helping to explain their acceptance or rejection of food aid. The Southern African countries were concerned about their export prospects with the EU if they accepted GM food aid in whole grain form. If GM food aid were planted and crossed with local varieties, this could affect exports of maize. Zambia, for example, exports some maize to European countries, and Zambia and other countries in the region did not want to close the door to potential future markets in Europe for GM-free maize exports (Institute for the Study of International Migration, 2004: 18–9).

Conclusion

It is unfortunate that the debate over biotechnology has been played out in the developing world through the politics of food aid. It has profoundly affected the recipient countries, and their environments and future trade prospects may suffer from it. The literature on food aid has until now paid insufficient attention to the question of GMOs and the impact they have on the food aid regime. I argue that it is time to insert the question of agricultural biotechnology squarely into the debates on food aid. The food aid regime is influenced by a number of factors that are unique to an age of agricultural biotechnology. These include the scientific debate over the safety of GMOs, and economic considerations linked to markets for GM crops. Both of these factors appear to have had an important influence on the policies on GM food aid pursued by both donors and recipients. In many ways, these factors are hard to separate from one another, and both are highly political.

The notion put forward in the early 1990s that the food aid regime had become largely 'depoliticized' must today be questioned. It is clear that the advent of agricultural biotechnology has fundamentally changed the nature of the regime.

Notes

1. I would like to thank Kate Turner, Sam Grey, and Christopher Rompré for research assistance, and the Social Sciences and Humanities Research Council of Canada for research support.
2. By late 2005 the following countries had adopted biosafety legislation: Cameroon, Ghana, Kenya, Lesotho, Malawi, Mauritius, South Africa, Swaziland, Tanzania, Uganda, and Zambia.
3. The Protocol speaks of living modified organisms (LMOs).

TP248 b5 . F16

(5 95 2001)

Part III The Transatlantic Divide and Its Global Impact

6
Biopolitics in the US: An Assessment

Kelly L. Kollman and Aseem Prakash

Background

Much has been written about regulatory divergence between the US and EU (Vogel, 2003; Vig and Faure, 2004). The case of biotechnology provides interesting insights regarding regulatory politics and how it has evolved in the US over the years. The EU has applied the 'precautionary principle' to regulate GM foods/agricultural biotechnology and has adopted a number of directives aimed at ensuring consumer safety. By contrast, the US has decided that GM products are no different from those made using more traditional methods. Consequently, the US government has neither enacted new statutes nor implemented new risk assessment procedures to regulate GM products. However, in recent years the US has shown signs of inching toward the EU's biotech policy mode in terms of the core issues of labelling and the segregation of GM and non-GM crops. How does one explain such changes in US biopolitics?

We draw on Kingdon's (1984) agenda-setting model to explain why the regulation of biotechnology has developed so differently in the two polities and how international pressures have come to affect, albeit subtly, the regulatory climate in the US. Kingdon argues that domestic policy agendas are shaped by three separate process streams: the political, problem and policy streams. The problem stream is determined by the extent to which the current situation in a particular issue area is perceived to be getting worse. The political stream includes interest group politics, the composition of the government and legislature as well as party politics. The policy stream is made up of possible solutions to perceived problems. Normally these three streams work in isolation from one another. However at certain moments a so-called 'policy window' opens – usually in response to a change in the problem or political streams – in which preferred policy solutions can be matched to policy problems. Using this general framework, the first part of the paper examines why, despite pressure from MNEs and NGOs for harmonized regulation (the former seek minimal regulation while the latter favor

103

ratcheting up standards), the regulation of agricultural biotechnology in the EU and the US has followed such different policy trajectories. We locate the explanatory variables in this section of the paper mostly within Kingdon's political stream or what we call the domestic political economy. We define the domestic political economy, our independent variable, as the relative strengths and relationships of actors who engage in the domestic policy debate about the use of GM products.

The second part of the paper examines why the US is showing signs of inching toward the EU's biotech policy mode in terms of the core issues of labelling and the segregation of GM and non-GM crops, thereby signifying a possible move toward a weak 'convergence to the top.' We provide evidence from developments at the federal level, state level as well as the changed US position toward a key international agreement, the Biosafety Protocol. We focus on the role of a critical 'domestic shock' (Sabatier and Jenkins-Smith, 1993; Baumgartner and Jones, 1993), the StarLink corn scandal. We argue that this event helped create the political space necessary for the US-based anti-biotech NGOs to change the tenor of the debate by significantly altering the problem stream. Once this 'policy window' was created, the biotech policy mode in the US began to respond to the pressures from foreign markets on US exporters as well as to the norms of consumer empowerment and transparency. As will be seen, the policy stream seems to be particularly susceptible to influences from the outside as MNEs seek to find common rules by which to operate in multiple national jurisdictions.

Biopolitics in the US

The US government has enacted no new statutes to regulate biotechnology products and has largely acted in concert with US-based MNEs to ensure minimum levels of regulatory hurdles. In part, this outcome was due to the Reagan–Bush era in which these decisions were made, that was marked by a belief that minimal regulatory oversight is necessary to foster US competitiveness in this critical emerging technology. In the mid-1980s, the Reagan administration rejected a proposal that would have made the Environmental Protection Agency (EPA) the lead regulatory agency for agriculture biotechnology products. In 1986, the White House issued an alternative scheme called the Coordinated Framework for the Regulation of Biotechnology that outlines the regulatory procedures still used today (Vogt and Parish, 1999). This scheme bestowed the agri-business-friendly USDA with the lead regulatory role.

Because US farmers increasingly rely on biotechnology for key crops such as corn, soybeans and cotton, the USDA has been a key proponent of agricultural biotechnology (Glickman, 1999). Within the USDA, the Animal and Plant Health Inspection Service (APHIS) is the lead agency for biotechnology regulation. Under the authority vested by the Federal Plant Pest Act, it approves field-testing of GM crops. Since 1987, APHIS has approved over 5000 field trials on about 28,000 sites (Foudin, 2000). Before any GM crop can be marketed, APHIS needs to approve a petition for a determination of 'non-regulated status,' that is, to certify that the crop is not a pest. This petition is published in the Federal Register that provides interested parties with the opportunity to express their views on how the GM crop may impact the environment and/or health safety of humans and animals. On this count, the approval process is quite open and subject to public scrutiny. Two other federal agencies, the Food and Drug Administration (FDA) and the EPA, are involved in regulating biotechnology. Under the authority vested by the Federal Food, Drug and Cosmetic Act, the FDA's Department of Health and Human Services regulates foods and food additives (except meat and poultry). In May 1992, the FDA issued a policy outline stating that GM and non-GM foods will not be treated as separate entities. Thus, unlike the EU, the FDA does not require labelling or pre-market safety studies for GM foods. As such the FDA is not deemed a critical regulatory agency for biotech products.

The EPA is involved in biotechnology regulation because it regulates pesticides under the Federal Insecticide, Fungicide, and Rodenticide Act and sets tolerance levels for pesticide residues in foods under the Federal Food Drug and Cosmetic Act. Because many GM crops contain pesticides, they require EPA's pre-market approval. Critics believe that this calls for long-term safety studies that are currently not required. Although never adopted, the EPA has proposed rules that would regulate pest-resistant GMO plants in the same manner as pesticides. In general, the EPA has voiced greater concern about the potential risks of the release of GMOs into the environment than the USDA or FDA; however, it has only rarely turned these concerns into concrete action.

The US regulatory approach differs markedly from the one adopted in Europe where most of the rules governing the regulation of agricultural GMOs have been written by the supranational European Union. In keeping with the common European perception that genetically modified plants are potential pollutants, the EU Commission's Environment Directorate General has become the lead agency responsible for drafting the horizontal legislation governing GMO regulation. As Council Directive 90/219/EEC (on contained use of live GMOs such as GM crops) and Council Directive 90/220/EEC (on field trials and marketing of live GMOs) mandate, GMO products must undergo a special application process to be used in field trials or to be put on the market in the EU. Drawing on the precautionary principle, which encourages preventive regulation even in the absence of scientific proof of harm, these procedures include fairly stringent risk assessment studies whose results must be sent to the Commission as well as to the governments of all member states before any GMO products can be introduced to the market.

In 2001, the Council adopted a revised version of the 90/200/EEC Directive which largely keeps in place the permitting procedures outlined above and augments them with several additional requirements. Starting in 2002, all marketing permits for GMOs will expire after a 10-year period at which time new applications and new risk assessments will have to be submitted if the product is to remain in the market. Additionally, national registers have been established that keep track of the locations of GMO plantings in the member states. The revised directive also increases the scope of monitoring that is done after a GMO variety has been approved for the market and outlines the circumstances under which permits can be revoked before the 10-year period of the permit has expired (EC, 2001). In addition to the permitting procedures, the EU has also established a fairly comprehensive GMO-labelling scheme that has no equivalent in the US. These rules set rather stringent thresholds by requiring that products containing 0.9 percent of approved GMO varieties and 0.5 percent for unapproved GMO varieties be labeled.

To sum up, the US has passed no new legislation for the regulation of GMOs and has largely adopted the opinion that products created by gene-altering processes are not fundamentally different from those made using more traditional means. The lead regulatory agency, the USDA, is well-known for being (agri)business-friendly. As such, the agency that was most likely to treat the regulation of biotechnology in an adversarial manner, the EPA, has played only a minor role in these regulatory decisions. In contrast, the EU has applied a strict interpretation of the precautionary principle. While American regulators have couched the biotech debate in the scientific language of risk assessment and cost–benefit analyses, their counterparts in Europe have emphasized a broader range of concerns such as food safety and the threat to traditional farming practices (Byrne, 2000b).

Explaining US biopolitics

Despite the susceptibility of biotechnology regulation to such globalizing pressures as international agriculture markets, promotion by large MNEs and protest from NGOs, the early phases of the policy debate, which occurred in the late 1980s and early 1990s, were largely structured by the domestic political economy surrounding the issue. Thus, differences in the kinds of groups involved in shaping the domestic agenda, the relative strengths of these groups, the issue linkages these groups were able to make as well as their relationships with regulatory authorities seem to be key to understanding why two such different regulatory modes developed in the US and the EU. An important determinant of the agenda-setting process in most democracies is public sentiment. Indeed the difference in public perceptions of biotechnology across the two shores of the Atlantic is stark. Although reliable longitudinal data is not available, multiple polls since the late 1990s suggest that Europeans are far less supportive of biotechnology than Americans. For example, Environics, an environmental polling firm, asked 1000 citizens in 25 countries about their views on using biotechnology to grow pest-resistant crops. Countrywide support levels for agriculture biotechnology were as follows: China: 79 percent, US: 78 percent, India: 76 percent, Germany: 54 percent, France: 52 percent, Britain: 36 percent and Spain: 29 percent (Washington Post, 1999). These findings are consistent with several non-comparative, longitudinal surveys carried out separately in the US and the EU. The International Food Information Council carried out annual surveys of US public opinion on GMOs from 1997 to 2002. In all of the surveys, a greater percentage of Americans reported a propensity to buy GM food than reported a reluctance to do so. However, the gap between supporters and detractors of GM food did diminish between 1997 and 2002 (Bonny, 2003). In Europe, the EU-commissioned Eurobarometer similarly has followed European attitudes toward GM food. Like in the US there has been a growing trend toward skepticism of GMO products. The difference, of course, is that Europeans have been much more hostile toward GMOs from the beginning. Since Eurobarometer began polling on this issue in the mid-1990s, they have recorded a strong desire by Europeans to avoid consuming GM foods. In the latest poll of 2001, over 70 percent of respondents voiced such a desire. Additionally, 56 percent of respondents believed that GM food was dangerous while only 17 percent rejected this idea (Bonny, 2003). We attribute the differences in public opinion to a variety of factors, especially differing economic structures and recent histories with food safety regulation that, in turn, led to varying institutional responses.

The most obvious difference can be found in the size of each region's biotech sector. According to a 2002 Ernst and Young survey of the 'life sciences sector' from 1994–2000, the growth of the US biotech industry has consistently outpaced that of its European counterparts. In 1994, there were less than 500 biotech firms in Europe and over 1300 in the US. While Europe had caught up with the US biotech sector in terms of firms by the late 1990s, the sector still lags behind its American counterpart in terms of employees, research and development investment and revenues. While the US has employed over 100,000 people in biotech jobs since 1994, Europe biotech companies still only employed 60,000 people in late 2001. In that same year, European biotech firms earned over 6 billion Euros in revenues – up from less than 1 billion in 1994 – but still lagged far behind US revenues of over 19 billion Euros. Similarly in 2001, US firms outspent its European competitors by a margin of 3:1 on research and development investment (Ernst and Young, 2002). By early 1999, the US had approved 40 GM crops for commercial marketing as opposed to nine approved by the EU with none being approved during the moratorium that lasted from 1998 until 2004.

Differences in the structure of the two regions' agricultural sectors also need to be taken into account. In the past four decades, the agriculture sector in the US has become increasingly concentrated and industrialized. It has emerged as a major exporter of agricultural commodities by virtue of becoming a low-cost supplier. Since 1975, export revenues have accounted for 20–30 percent of US farm income (USDA, 2001). Over the past decade, US agriculture exports have made up about 13–14 percent of the world's total agricultural exports. With growing competition from China and Latin America, US agriculture producers have had to work hard to maintain this position. One reason for this success is its reliance on large-scale modern technology, including biotechnology. Approximately, 61 percent of cotton, 87 percent of soybean and 45 percent of corn grown in the US use GM seeds. Acreage under GM crops has grown from 6 million in 1996 to just over 105 million in 2003 (USDA, 2005). In 2004, the US planted 59 percent of the world's GM crops. In the EU, Spain is the only country that grows a significant number of GM crops. Its total GM acreage, however, amounts to less than 1 percent of the world total (James, 2004). Additionally, to a greater extent than is true in the US, the family farm in Europe has consciously been protected against the potentially devastating effects of the world market under the auspices of the EU's Common Agricultural Policy. Europeans have come to justify these protectionist practices which exist at both the national and EU levels for cultural reasons and out of a concern for the darker side of the mass production techniques (such as biotechnology) used in the US (Byrne, 2000a). It is thus not surprising that many of the trade disputes that have arisen between the US and the EU in the past decade have in some way involved agricultural issues.

Of course, this begs an explanation for US opposition to labelling even outside the United States. If US firms were major exporters of GM crops to the EU, this opposition could be understood. With the exception of soybeans, the major markets for US agriculture exports lie outside the EU. Thus, US opposition so far can be explained by the fear that there may be a convergence-to-the-top instead of a race-to-the-bottom, whereby EU GMO-labelling standards may be accepted as *de facto* standards by countries outside Europe as well. As we discuss subsequently, economic globalization in the form of pressures from importing markets is creating incentives for many US exporters to demand some sort of verifiable methods of segregating GM and non-GM crops.

This is not to say that European MNEs are shunning biotechnology. Biotechnology has been extensively used in the pharmaceutical industry by European MNEs. Since biotechnology applications in pharmaceuticals have yet to be debated vigorously, this industry is not being scrutinized or opposed on this count. In fact, European pharmaceutical firms have conspicuously maintained silence on the raging debate over the morality and usefulness of biotechnology lest they become the next targets of anti-biotechnology groups.

Recent food safety disasters in Europe have accentuated the European public's discomfort with GM foods. Although the European public was quite skeptical about their food regulatory institutions before the recent outbreaks of BSE mad cow and foot and mouth diseases (Pollack and Shaffer, 2001), these incidents have been a key driver in the recent revisions of European biotech regulation that have ratcheted up standards further. This mistrust is accentuated by the institutional deficit in terms of a lack of independent regulatory agencies such as the EPA, FDA or the USDA in the EU or most member states. Because European countries have historically tended to regulate core industries through nationalizations, independent regulatory agencies have never been used extensively. With the privatization of most of these state-owned enterprises, European governments have increasingly turned to independent agencies to enhance government oversight capabilities (Majone, 1996). In the wake of the recent food safety catastrophes, the EU has decided to follow this trend and created the new independent European Food Safety Authority (ENDS Daily, 1999).

To sum up, the weakness of the biotech industry, the cultural importance placed on small-scale farming and recent food safety disasters have greatly affected the political economy surrounding the biotech industry in Europe. Because consumers in Europe have come to view GMOs as potential pollutants, the EU's Environment Directorate General logically became the lead regulatory body on this issue. This strong public reaction and its antecedent causes go a long way in explaining EU's aggressive approach to regulating GMOs. In the US, by contrast, the strong presence of the biotech industry, a farming sector focused on price competition and exports and a public that is not particularly focused on food safety concerns has caused biotech to remain a largely technocratic issue regulated primarily by the industry-friendly USDA.

Biopolitics toward convergence?

Normally, our examination would end here. But starting in 1999, there has been a subtle but noticeable change in the policy agenda surrounding biotechnology issues in the US, signifying movement (but not a drastic change) toward some sort of convergence with EU standards, especially regarding the labelling of GM crops. We posit that international pressure is beginning to have an effect on the regulatory climate for agricultural biotech products in the US. However, these international influences do not work in a straightforward manner and are always filtered through Kingdon's domestic agenda streams, the political stream (our domestic political economy variable) and the problem and policy streams.

What exactly has changed in the US regulatory climate since 1999? We argue that three key changes have occurred in the US approach to regulating agricultural GMOs in the past six years: (1) an increase in the number of court cases challenging the marketing and release of GMOs into the environment; (2) the introduction of bills in both Congress and state legislatures calling for more stringent regulation of GMOs; and (3) the USDA's and the

FDA's review of their GMO policies and their about-face on recommending labelling for GMO products. Although these changes do not represent revolutionary shifts in policy, taken together they do signify a trend toward a transformation of the regulatory climate.

The first change has been the mushrooming of legislative activity that has occurred since 1999 at both the federal and especially the state levels. Many bills have been introduced in the US Congress and state legislatures that call for more stringent regulations and mandatory labelling. The surge in legislative activity began in the 106th Congress with the introduction of several bills. This interest continued in the 107th Congress with the introduction of 50 bills.[1]

There has been significant legislative activity at the state level as well. In 2000, 13 bills were introduced in various state legislatures. In 2001, 130 pieces of legislation were introduced in 36 states. Ninety-three of these bills focus on such issues as regulation of GM crops, including bans or moratoriums on their use. Maryland has barred the introduction of GM crops while North Dakota has barred the introduction of GM wheat for 2 years. In addition, bills were introduced in several states to place a moratorium on GM crops. These efforts include proposals in Massachusetts, Montana, New York, South Dakota and Vermont (Pew Initiative on Food and Biotechnology, 2002). Twenty-two pieces of legislation were introduced in 10 other states (Colorado, Hawaii, Massachusetts, Michigan, New Hampshire, New York, Oregon, Rhodes Island, Vermont and West Virginia) for either the voluntary or mandatory labelling of all food products generated through biotechnology. North Dakota passed a bill requiring the segregation of GM crops from non-GM crops. In addition, 22 bills were introduced in Hawaii, Massachusetts, Maine, Minnesota, North Carolina, Oregon and South Dakota requiring some sort of restrictions on the sale and use of GM crops (Pew Initiative on Food and Biotechnology, 2002).

The pace of regulatory legislative activity slowed somewhat in 2003 and 2004, in part because the desired legislation had been passed in many of these states. At the same time the number of bills proposed at the state level that support GM food has increased slightly since 2003. These latter bills mostly fund research on biotech agriculture or protect farmers from the destruction of GM crops. Still important regulatory measures have been adopted over the past 2 years by a number of state legislatures which includes eight bills regulating the release and distribution of GMOs, one liability bill and three bills introducing segregation and labelling schemes for certain GMO products (Pew Initiative on Food and Biotechnology, 2005). Legislative debates over how a new variety of 'RoundUp Ready' wheat should be regulated in the northern plains states of Montana and the Dakotas resulted in Monsanto delaying, perhaps permanently, its plans to market this new product (Pew Initiative on Food and Biotechnology, 2005).

This mushrooming of legislative activity at the state level has resulted in a patchwork regulatory system in the US. States in the northeast part of the country, which have smaller-scale farming and are heavily invested in the organic food sector, have passed quite stringent legislation including strict labelling schemes and bans on certain kinds of GM varieties. In the Midwest, where many GM foods such as soybeans and maize are grown, legislatures have sought to pass laws to help local farmers benefit from GM products where they can and to protect them from anti-GM food sentiment where it is strong. These legislatures thus support research of GM agriculture applications but have sought to introduce liability laws to shield individual farmers from potential losses associated with planting GM crops. These states have also reacted strongly against the proposed introduction of Roundup Ready wheat. Hawaii, which still relies heavily on commercial agriculture, has sought to create a more comprehensive regulatory structure. Like the Midwestern states, its package of laws tries to take advantage of GM food markets where it can and protect producers against backlash in more controversial markets. It has recently imposed a moratorium on transgenetic fish. Since California recently followed its lead and also placed a ban on this GM food, the future of the transgenetic fish market seems very much in doubt (Pew Initiative on Food and Biotechnology, 2005).

At present the US Congress seems to be content to let states work out individual regulatory schemes. Very few bills either regulating or prohibiting the regulation of GM foods have been passed at the national level and fewer have been introduced in the 108th and 109th Congresses. While this state-led activity has resulted in a much less comprehensive and less stringent regulatory system than exists in Europe, its consequences are not insignificant and have led to a rudimentary segregation and labelling scheme in certain US states.

The second major change to occur in the US has taken place in the judicial arena as biopolitics appears to be no longer immune from the adversarial legalism that is so prominent in the environmental policy arena. In early 1999, the Center for Food Safety (2000) initiated legal action against the FDA demanding that Monsanto's genetically engineered Bovine Growth Hormone be taken off the market. The CFS suit was prompted by Health Canada's report (the Canadian equivalent of the USFDA) that the FDA had failed to investigate studies that indicated that the oral feeding of the growth hormone led to a 25 percent risk increase of mastitis (udder infections) in dairy cows.

In December 1999, a coalition of small farmers and farm groups filed a class action suit against Monsanto accusing the company of marketing GM seeds without properly testing them and giving farmers false guarantees about the marketability of GM crops. Further, Monsanto and nine other companies are accused of forming an international cartel that has conspired to control the world's market in corn and soybean seeds (CMHT et al., 2001).

In December 2000, a class action suit on behalf of US farmers was filed against Aventis alleging that the company failed to take precautions to ensure that StarLink does not enter the human food supply chain (CMHT et al., 2001). In a 1999 report on the future prospects of the agriculture biotech sector, Deutsche Bank warned that the industry could likely be saddled with law suits in numerous jurisdictions if the problem of unintentional GMO contamination of adjacent fields and grain supplies is not resolved (Guardian, 1999). Several such court cases are already beginning to appear in US districts courts (Moeller and Sligh, 2004).

Still the anti-biotechnology coalition has refrained from introducing the kind of legal adversarialism into the policy-making process that marked the early court battles fought over environmental policy issues. While biotechnology for the first time is being portrayed as a new generation of pollution and a public health issue, most groups are asking for rigorous screening and safety protocols rather than outright bans on the use of biotech products. The reason, ironically, lies in the short-term perspective that industry has adopted. Because of their huge investments in research and development, biotechnology companies have been pushy in seeking to market their products. Consequently, they have been less willing to invest in long-term impact studies, field trials and farmer education. Thus, framing the debate in terms of benefits and costs makes the anti-biotechnology alliance look reasonable and puts business on the defensive. Second, the anti-biotechnology movement has come to realize that although the American public is becoming concerned about biotechnology, the problem has not been sufficiently radicalized to ignore economic implications of restrictive regulations.

The third change has occurred within US regulatory agencies themselves which have begun to reexamine their regulatory deficiencies particularly in the area of labelling and carrying out additional risk assessment procedures. The USDA has acknowledged that although it is supposed to be an independent regulatory agency, its promotional functions may clash with its regulatory functions. Thus, to remove any accusation of conflict of interest, former Secretary Glickman (1999) announced a review to ensure that the USDA maintains a credible regulatory process. In May 2000, the Clinton Administration declared (and the Bush Administration has not rescinded) that the USDA, the FDA and the EPA would review agricultural biotechnology regulations (White House, 2000). In response, the FDA published draft rules in January 2001 that require firms to notify the FDA 120 days in advance before they market GM foods (*Federal Register*, 2001) and encourage firms to voluntarily consult the FDA. They also specify the safety test data that should accompany the notification (Vogt and Parish, 2001). In 2004, the FDA issued a new set of guidelines that recommended manufacturers carry out early, pre-market food safety evaluations of new bioengineered proteins. These new recommendations are meant to ease consumer fears about the possibility of GMOs entering the food supply chain that are approved for field trials but not yet as food additives. Despite this tightening of the rules, the FDA still does not require pre-market approval of GMO foods although it does strongly recommend that manufacturers take part in their 'voluntary, pre-market consultation' program.

After opposing the labelling of GM foods for almost a decade, beginning October 2002, under the USDA's National Organic Program, three types of organic labels have been introduced: '100 percent organic,' 'organic' or 'made with organic.' The use of GMO ingredient is prohibited for any kind of organic label (USDA, 2003). This is an important policy shift given that USDA has maintained that GM crops are no different from non-GM crops. This change is not surprising because consumer studies commissioned by the FDA (2000) suggest that US consumers are becoming anxious about the long-term impact of GM foods. A poll conducted by the Pew Initiative on Food and Biotechnology in January 2001 indicates that about 75 percent Americans want to know if their food has been genetically altered (MSNBC, 2001). Following a study by the National Research Council[2] commissioned by the USDA, the agency announced its intention to revamp its GMO permitting and market application regulations in January 2004. The new rules, which are not yet in force, would likely expand the kinds of GMO crops the USDA can regulate as well as allow the USDA to continue to regulate certain high-risk GMOs after they have been approved for market use.[3]

While events in the international arena have influenced US domestic actors' stances toward GMO regulation, this influence always works through the prism of domestic political institutions, processes and past policy legacies. It is for this reason that Kingdon's agenda-setting model is still appropriate for examining the policy outcomes of regulatory fields that are shaped by international rule-making. We begin by examining international factors. The pressure to ratchet up GMO regulation was first felt by US MNEs seeking to gain access to stringently regulated foreign agriculture markets, especially those in Europe.[4] By the late 1990s, it was becoming increasingly clear that consumers in key US agriculture export markets were extremely distrustful of GMO products and the US MNEs that market them. A number of well-publicized attacks on fields where GMO crops had been planted in Europe and India did a great deal to publicize the issue to consumers across the globe (Depledge, 2000: 160–1).

Consumers in Europe and parts of Asia began insisting that US MNEs be required to segregate and label GM seeds and foods so that they could choose to avoid them if they wished. These calls for labelling have been translated into concrete legislation in a growing number of countries. By June 2001, governments in most European countries, Brazil, China, India, Australia, New Zealand, Russia, Israel, Ecuador, Paraguay, Thailand, Indonesia, Hong Kong, Pakistan, Indonesia, Egypt, Algeria, Saudi Arabia, Sri Lanka, South Africa, Ethiopia and Ghana had adopted mandatory labelling and other types of restrictions on GMO agriculture products (OCA, 2001).

The pressure generated by this consumer protest and the moves toward mandatory labelling abroad was first felt by the US MNEs that were often the targets of these protests, particularly Monsanto. As the manufacturer of GM corn, soybeans and cotton, Monsanto has aggressively marketed GM food crops across the globe throughout the 1990s usually by ignoring requests for the segregation and labelling of GM foods and seeds. These tactics have been widely criticized by environmental and consumer groups in Europe and Asia. In response to this consumer backlash, a number of large firms in Europe including McDonalds and McCain Foods have asked their contract farmers to produce non-GM crops. As a result, Monsanto eventually withdrew GM potatoes from the market. These developments not only hurt the sales of biotech MNEs but it also influenced their stock prices. In 1999, the value of Monsanto's shares fell by 25 percent (Chase, 2000: 3D).

These dramatic events throughout 1999, mainly induced by NGO pressure, caused Monsanto and other US MNEs to change their positions on establishing an international agreement regulating the trade and transport of GMOs. Talks for such an agreement, a Biosafety Protocol to the Convention on Biological Diversity (CBD), had been under way since the mid-1990s. After initially opposing such a treaty in the 1980s, Monsanto paid lip service to supporting these negotiations in the 1990s but always pushed for a very weak agreement. Its influence was felt at the negotiations in Cartagena, Columbia in early 1999. Although the US did not ratify the CBD, it actively participated in the negotiations as an observer and through its five allies that have signed the CBD. The US blocked a proposed treaty that would have required exporters of GM products to obtain advance permission to distribute these products from importing countries and that GM products be segregated and labeled (Pollack, 2000).

By the time negotiations for the Biosafety Protocol resumed in January 2000 in Montreal, both the US negotiators and Monsanto were willing to agree to a much stronger treaty than was the case just a year before. As outlined above, the increase in consumer protest against GMOs and corresponding action by large restaurant retailers had caused Monsanto to change its decade-long opposition to any form of labelling as well as the use of the precautionary principle in regulating GMOs. The agreement that was signed in Montreal, which became known as the Cartagena Protocol on Biosafety, requires shipments of GM crops to be identified and allows importing countries to reject these shipments based on the precautionary principle. Although US negotiators allowed the inclusion of the precautionary principle, they also insisted that import bans that are not based on scientific data are susceptible to WTO rules against discrimination (Alden, 2000: 11). Thus, where the line is to be drawn between precaution and discrimination remains unclear. Still this agreement, which was praised by both Monsanto and Greenpeace, signifies a major shift in US trade policy and for the first time acknowledges the legitimacy of EU labelling standards and use of the precautionary principle in the regulation of GM products. As mentioned above, however, the US is not a signatory to the CBD or any of its Protocols. Unlike the signatory countries, therefore, the US is not legally bound to implement the rules in the Biosafety Protocol.

Despite this activity at the international level, very little change occurred in the US domestic arena until late 2000. Indeed the most critical change occurred due to an unexpected development – the StarLink corn episode – that opened up a 'policy window,' thereby creating opportunities for the anti-GM coalition to redefine the of biotechnology issue within the US context. This incident took place in September 2000.[5] A laboratory testing on behalf of a coalition of anti-biotech groups discovered that taco shells sold under the Taco Bell Shells label of Kraft Foods contained traces of GM corn. This particular corn, StarLink, produced by Aventis Crop Science, had been approved by the EPA for animal use but not for human consumption.[6] Faced with a public outcry, Kraft recalled all taco shells. Within two weeks of this incident, Safeway, a major US grocery chain, found StarLink in its house brand of taco shells. At the urging of the EPA, Aventis 'voluntarily' canceled its marketing license for StarLink. Only after this domestic shock occurred did the problem stream in the US begin to change significantly. Stories on biotech regulation began to appear regularly in *The New York Times*, *The Washington Post*, *USA Today* and even on the CBS Evening News. A standard search on LexisNexis found that 81 articles were published about GMO–biotechnology in the year 2000, 33 articles in 1999, 8 articles in 1998 while no articles were found for 1997 or 1996. Additionally, beginning in 1999, well-resourced consumer and environmental groups have started to pay more attention to biotech issues and questioning the US's comparatively lax regulatory regime since. Prior to that, the anti-biotechnology crusade in the US was led by groups/activists such as Jeremy Rifkin who were generally perceived as being extremist and their arguments lacking scientific validity.

This change in the problem stream, which was only tangentially touched off by events occurring outside US borders, has opened up the political space for a wider debate about GMO regulation. It is here that we see the very obvious influence of international rule-making. In particular, the policy stream, which seeks to match solutions to problems, has been greatly shaped by the rules and regulatory principles adopted at the international level and by foreign governments. The legislative, judicial and inter-agency debates that have opened up in the US since the StarLink episode have centered around the core issues of the segregation of GM products and labelling and, to a lesser extent, the use of the precautionary principle. The hesitant moves made toward the European model show that the political stream in the US is still greatly influenced by a strong biotech industry and an agriculture sector geared toward price competition.

Conclusions

What does the GMO case tell us about regulatory politics? Is the US convergence to EU standards weak or strong? To assess these issues, one should examine the extent to which either party's 'core positions' on the definition of a policy problem, the use of certain policy instruments and the institutional structures used to address the problem have changed. High convergence implies changed core positions. In biotechnology, the core issues are labelling and segregation. At this point, convergence is weak because neither party has formally compromised on its core position. However, we find weak convergence to EU standards because some US states have enacted labelling laws, the Federal government has watered down its opposition to labelling and there was a surprising shift in the US position on the Biosafety Protocol.[7] Thus as we argue elsewhere, policy convergence should be understood as the movement of two or more societies toward a common position; it is not the process by which two polities come to have the exact same policy outcomes. The US and the EU do not have the same regulations in place to regulate GM agriculture products. Rather the US has made some partial movements toward the European model (Prakash and Kollman, 2003).

The EU and US comparisons are interesting precisely because examples can be found from across the divergence–convergence continuum. The regulatory dispute over beef hormones is characterized by continued divergence even in the face of a WTO ruling against the EU. By contrast, the successful negotiation of the worldwide ban on most ozone-depleting substances is a case of strong convergence between an initially reluctant EU toward a more enthusiastic US. The cases of GMO regulation, fisheries management and mining in the Antarctic represent cases of weaker convergence. Our argument suggests that these differences across cases can only be explained by examining how international pressure for change is filtered through and sometimes subverted by the domestic political economy. When upward convergent change has occurred between the US and the EU in the environmental sphere, it has tended to be of the 'weak ecological modernization' variety. There has been a ratcheting up of rules but not to the highest standards and generally without dramatically changing how the core issues are being defined in individual polities.

While our findings go some way in pinpointing the circumstances under which international pressure causes domestic policy change as well as the extent of that change, more research in this area is needed. To the extent that upward convergence was achieved in the regulation of GMOs, it occurred in Kingdon's policy stream in which policy instruments are matched to policy problems. The US government is likely to make concessions to the EU on labelling without changing how the problem of GMO regulation is defined in the US and without changing the institutional structures erected to deal with GMOs in the US. Given the need for common rules in international markets, it seems quite logical that policy instruments would be more vulnerable to change than domestic policy institutions or core aspects of how an issue is framed. This hypothesis should be tested in other cases.

To conclude, upward regulatory convergence may be possible particularly in instances where NGOs can successfully put pressure on MNEs, but this process is seldom complete. Convergence, either in terms of races to the bottom or to the top, is not inevitable; the domestic political economy can sustain both regulatory divergence as well as help facilitate increased convergence. National policy agendas surrounding biotechnology regulation have been influenced by international forces but only to the extent that the domestic problem streams have made room for these issues and only in a manner dictated by domestic politics. What political scientists have been slower to recognize is the more direct effect that NGOs can have on MNE policies and behavior. Along with studying the influence of NGOs on state policy, political scientists should also focus on how they influence MNEs directly.

Notes

1. For a detailed listing of the bills, see Thomas, The Library of Congress (2000) http://thomas.loc.gov/, accessed 19 August 2002.
2. http://books.nap.edu/catalog/10880.html?opi_newsdoc01202004, retrieved August 5, 2005.
3. http://pubs.acs.org/subscribe/journals/esthag-w/2004/jan/science/rp_usda.html, retrieved August 5, 2005.
4. For an account of how international market pressure has affected US biotech regulation see Young (2003).
5. We do not consider the Monarch butterfly episode to have had the same impact, although it received a lot of media coverage. For one, it did not result in the recall of food products or other mass-produced items. Further, the Monarch butterfly study has been extensively criticized on methodological grounds.
6. Subsequent to StarLink episode, Aventis asked the EPA to approve StarLink for human consumption but, much to the dismay of the food industry, was turned down (Kaufman, 2001).
7. In the beef hormone case, there is continued divergence. The core EU position is that it will not allow the import of hormone-treated beef – labelling is not a core issue for the EU. In the face of US pressure, if the EU allows import of labeled hormone-treated beef, then the EU would have compromised its core position, and the convergence would be high and toward US standards.

7

The Transatlantic Agbiotech Conflict as a Problem and Opportunity for EU Regulatory Policies

Les Levidow

Introduction: explaining the transatlantic conflict

The US and EU have developed quite different frameworks for agbiotech regulation, but these were potentially compatible with transatlantic trade in GM products. How, then, did a trade conflict arise? And why is it a problem?

Since the early 1990s the US government policy has treated agbiotech as technological progress yielding beneficial products which warrant no special label, and whose safety can be readily demonstrated through 'sound science'. Federal agencies readily approved or deregulated GM crops for cultivation and other uses. Food regulators routinely accepted company claims that GM foods had substantial equivalence with a safe counterpart, as grounds not to require safety approval.

By contrast the EU system has based its regulatory system on uncertainty about risks. By enacting the 1990 Deliberate Release Directive, the EU required a risk assessment and safety approval for all GM products. From 1996–97 the EU began to approve some GM products already commercialised in the US. And the EU eventually required statutory labelling, which became successively more stringent.

After public protest mounted in the late 1990s, the EU regulatory procedure underwent delays and blockages. In 1999 the EU Council suspended the decision-making procedure for new GM products, and some member states banned GM products which had already gained EU-wide approval. A trade conflict with the USA led to anticipation and threats of a WTO case, which was launched in 2003.

Transatlantic regulatory divergence has been widely seen as the main source of the US–EU trade conflict. Moreover, each jurisdiction has been stereotyped in ways which diagnose why the divergence is a problem. According to some commentators, for example, the US bases its regulation on 'sound science', while EU restrictions and delays have accommodated the fears of an irrational public. According to other commentators, the US ʻ government bases safety claims on scientific ignorance and force-feeds the

118

world with GM food, while the EU defends precaution and democratic sovereignty. Although these arguments differ greatly in standpoint, they all emphasise transatlantic differences and consequent problems.

This chapter challenges such explanations centering on transatlantic regulatory divergence. Instead it will ask: How did the trade conflict arise from policy agendas which span the Atlantic? How has it been framed as a problem and used as an opportunity for policy agendas within the EU? (For analogous roles within US policy debate, see Murphy and Levidow, 2006.)

The rest of this Introduction summarises the overall argument as follows: The US–EU conflict arose from contending transatlantic agendas, which have operated within and across the two jurisdictions. Corresponding to each agenda, transatlantic coalitions framed the policy problem in three different ways: first, regulatory harmonisation for trade liberalisation of a benign technology and its safe products; second, the consumer right to know and choose safe food, based on precautionary risk assessment; and third, civil society participation in broad evaluation criteria for agbiotech products, which need prior proof of their safety.

The trade liberalisation agenda set the context in which European protest could frame agbiotech as a dual threat of 'globalization' and unknown risks. Greater controversy led to regulatory blockages and a trade conflict, which policy actors diagnosed in ways convenient for their own agendas. Promoting agbiotech, some politicians warned that EU regulatory delays or more stringent rules would be found illegal at the WTO. But this strategy backfired, instead providing a vulnerable target for attack by 'anti-globalization' activists. Citing US threats of a WTO case, opponents sought to delegitimise pro-agbiotech policies as a surrender to political and commercial pressures.

From the late 1990s onwards, some European policymakers articulated a new problem – how 'to restore public and market confidence' – as an imperative for institutional reform. This problem-diagnosis helped to bypass earlier disagreements about the 'scientific' basis of regulatory criteria, thus facilitating more precautionary approaches to risk assessment and GM labelling. These changes accommodated key aspects of the 'consumer rights' agenda, thus potentially establishing a stronger basis to legitimise EU decisions.

As the conclusion will explain, the overall analysis extends insights from other academic accounts (Bernauer, 2003; Jasanoff, 2005; Toke, 2004; Isaac, 2002). This chapter also disagrees with some accounts, especially those which attribute the US–EU conflict to distinct jurisdictional characteristics. That analysis draws upon the results of three research projects.[1]

As its structure, this chapter first sketches how contending transatlantic agendas generated the conflict. Then it shows how an EU regulatory impasse stimulated policy change in regulatory criteria and expert advisory arrangements. Finally the conclusion summarises how contending transatlantic agendas operate within EU agbiotech regulation and drive its ongoing tensions.

Contending transatlantic policy agendas

A trade liberalisation agenda set the context in which European protest could frame agbiotech as a dual threat of 'globalization' and unknown risks. Conflicts arose in the mid-1990s, especially around proposals to approve GM crops for cultivation, potentially leading to a US–EU trade conflict. This section describes how various policy actors used the conflict to elaborate and promote their contending agendas.

Using the transatlantic conflict: Bt maize crisis

As the EU statutory framework for agbiotech, the 1990 EC Deliberate Release Directive encompassed diverse agendas. It aimed to harmonise regulatory criteria for GM products, as a means 'to complete the internal market'. From this perspective, the policy problem was divergent national criteria which could impede internal trade. The solution lay in an EU-wide regulatory framework basing the internal market on a high level of protection for human health and the environment (EEC, 1990). By the mid-1990s those aims were linked with another policy framework: to promote agbiotech for European economic competitiveness (Levidow et al., 1996). These various agendas soon collided in decisions about specific GM products.

In late 1996 the EU Council members could not agree on whether to approve Ciba-Geigy's Bt-176 maize, present in grain shipments imminently arriving from the USA. The European Commission had the authority to make the decision itself, but its members could not agree, according to leaked minutes. The Trade Commissioner Leon Brittan argued that indecision in Brussels might anger the US government. Other Commissioners successfully advocated waiting for the opinions of three EU-level scientific committees, given the wider expert disagreements over safety. There were also disagreements about whether commercial approval should include a labelling requirement. Along with Leon Brittan, the Industry Commissioner Martin Bangemann opposed mandatory labelling; they argued that such a requirement might be illegal under international trade rules and could draw the EU into a trade dispute at the WTO (Rich, 1997: 8).

Meanwhile the mass media highlighted expert disagreements over product safety. Some experts were concerned that the antibiotic-resistance gene in the Bt-176 maize might enter pathogenic microbes and jeopardise the clinical use of the corresponding antibiotic. The Bureau Européen des Unions de Consommateurs (BEUC) emphasised several risks that had already been raised by some member states as a basis to oppose approval. But eventually the EU's expert bodies rejected all safety concerns about the product. On those grounds, the European Commission approved Ciba-Geigy's Bt-176 maize in January 1997, despite opposition from nearly all EU member states.

The decision was widely attacked as illegitimate, especially by analogy to the 1996 BSE crisis. When the European Commission's minutes were leaked in the Belgian newspaper *Le Soir*, it used the headline 'After mad cow, recidivism with transgenic maize' (Rich, 1997: 1). According to a Green Member of the European Parliament, 'Despite mad cow, they have learned nothing!' The Pesticides Action Network argued, 'This is crazy. They have started a gigantic experiment with us as the guinea pigs' (ibid: 8). In April 1997 the European Parliament voted overwhelmingly to denounce the European Commission for its approval decision.

According to the Commission minutes, US maize shipments had been creating a strong pressure on the European Union to approve Bt-176 maize as quickly as possible. Some Commissioners were anticipating a compensation claim from the US if they failed to authorise the product. However, the Consumer Affairs Commissioner, Emma Bonino, expressed regret that the approval decision was responding to economic pressures. She believed that the European Commission should reflect on consumer concerns and their desire for transparency. Mentioning the BSE crisis, Commissioner Neil Kinnock said that consumer confidence must be re-established; maize is widely used and GM maize would be difficult to identify in derived products (ibid).

In all those ways, an incipient trade conflict was being used for contending policy agendas within the EU. Arguments to minimise regulatory delays and criteria came from Commissioner Brittan, who was already championing trade liberalisation, and from Bangemann, who was promoting agbiotech as an imperative for the EU's economic competitiveness. Now they framed the trade conflict as EU delinquency which would result in a guilty verdict at the WTO. By contrast, other Commissioners echoed public concerns about scientific uncertainty and consumer choice. Some politicians sought to bypass the conflict through official expert advice, yet this too became part of the risk controversy. Meanwhile environmental NGOs sought to delegitimise any product approval, thus increasing the pressure upon governments to delay or reject such a decision. These policy agendas roughly correspond to formal coalitions, as described next.

Contending transatlantic policy agendas

Pressures for rapid EU approval came from a trade liberalisation agenda. In 1995 the Commission's New Transatlantic Agenda (NTA) identified 'barriers to transatlantic trade' as the main problem for EU and US policy makers. Representing multinational companies, the Transatlantic Business Dialogue (TABD) helped to define the NTA's aims and the overall policy agenda. Regulatory harmonisation was expressed with the slogan, 'Approved Once, Accepted Everywhere', at least at the transatlantic level, ideally leading to a New Transatlantic Marketplace. In 1998 the US and EU governments established the Transatlantic Economic Partnership (TEP), a quasi-technical government–business network, to help implement TABD proposals. The DG-Trade Commissioner Leon Brittan led the EU-wide promotion of the trade liberalisation policy (Murphy and Levidow, 2006).

For the agricultural biotechnology sector, TABD members identified pre-market safety assessment as the only regulatory issue. TABD emphasised the need for a common approach across the Atlantic, and ideally a centralised approval procedure, based on 'sound science'. Designed as a largely technical body, the TEP aimed to identify regulatory differences as a step towards overcoming them.

According to Isaac (2002), the NTA-TEP and TABD have both sought multilateral integration, towards 'allowing decentralised markets to achieve efficient and optimal objectives', though with different strategies. In his view, the TABD seeks to identify and coordinate any divergent regulations which may impede trade, and thus to develop a common regulatory framework. By contrast, the TEP seeks to internalise traditionally non-market, social objectives into such a framework (ibid: 27). Regardless of any such differences in emphasis, they shared a common agenda: how to harmonise risk assessment for trade liberalisation.

Although EU regulation of agbiotech diverged from the US model, the EC Deliberate Release Directive was foreseen as a means towards trade liberalisation. From the start, EU agbiotech regulation framed GM techniques as a novel process which warranted special risk-assessment procedures for all GM products. European Commission staff advocated this regulatory framework as a basis for harmonising EU-wide regulatory criteria (Levidow et al., 1996), even for overcoming transatlantic differences (Jasanoff, 2005: 82).

In the mid-1990s civil society protests began to challenge the neoliberal 'free trade' agenda, through movements widely called 'anti-capitalist' or 'anti-globalization'. Such protests created the context in which mainstream NGOs could more effectively challenge the NTA-TEP. Among other critics, consumer groups attacked the NTA-TEP for favouring industry influence and for 'levelling down' standards. Anti-agbiotech activists linked several threats: trade liberalisation policies, economic globalization, corporate power and unknown risks of GM products.

Facing legitimacy problems, in 1998 the US and EU invited NGOs to establish their own 'transatlantic dialogues', as a basis to participate in the NTA-TEP process. Each network devised a different strategic response to that general process and to agbiotech in particular. Through the Transatlantic Consumer Dialogue (TACD), consumer NGOs found shared goals in 'consumer rights', such as the 'right to safe products', the 'right to know and right to choose', and jurisdictional sovereignty to accommodate diverse standards from consumer demands. They demanded precautionary regulation and full labelling of GM food products. Relative to TACD members, environmental groups were more antagonistic towards transatlantic trade liberalisation and agbiotech. The Transatlantic Environmental Dialogue

(TAED) opposed agbiotech as long as there was no convincing evidence about its 'harmlessness to man and nature'. They proposed civil society participation in broad evaluation criteria for agbiotech products. They also opposed the concentration of corporate control over the food system (Murphy and Levidow, 2006).

Thus agbiotech was framed in three different ways, corresponding to the three transatlantic Dialogues. Proponents linked the technology to efficient agriculture, economic growth, and benefits to farmers and the environment. Mainstream consumer groups did not oppose biotechnology in principle but demanded stronger regulatory frameworks from a consumer rights perspective. Especially in Europe, environmental groups framed the technology as an ominous symbol of corporate domination, economic globalization and unsustainable agriculture.

From NGOs and some governments, critical voices gained a greater hearing in the EU policy system as intense public controversy erupted over agbiotech in the late 1990s. Protests coincided with a wider crisis of the agri-food safety system. The 'mad cow' epidemic was framed as a threat of intensive agriculture generating health hazards which elude the available scientific knowledge and official expert advice. According to some critics, intensive agriculture threatened alternative values and futures of European agriculture (Levidow and Marris, 2001). GM products were likewise turned into a symbol of such threats; EU democratic sovereignty was counterposed to 'globalization' and the TABD-TEP agenda. Thus the US–EU agbiotech conflict arose from contending policy agendas operating across the Atlantic.

Regulatory impasse as stimulus

In the late 1990s EU-wide regulatory conflicts challenged the narrow criteria which had facilitated product approvals. Beyond the Bt maize mentioned above, in 1997–98 the Commission granted EU-wide approval to some GM crops as normal commercial products, that is, with no requirement for special control measures. Such decisions provoked dissent from member states and became targets for anti-agbiotech critics, for example, on grounds that relevant uncertainties were ignored (Levidow, Carr and Wield, 2000).

Citing the familiar 'pesticide treadmill' as an analogy, agbiotech critics warned that GM crops would create a 'genetic treadmill', whereby weed or insect pests develop resistance to newly inserted genes. This resistance could pose a problem not only for pest control but also for additional pesticide sprays to control resistant pests. This scenario became a salient issue for GM herbicide-tolerant oilseed rape, which France prohibited in 1997. The debate there was used to generate resources for risk research; French scientists tested how far herbicide-tolerance genes flowed and persisted over several generations. The empirical results were cited to justify the original rationale for the French ban.

Potential harm to farmland biodiversity became a salient issue in the UK. Controversy focused on whether broad-spectrum herbicide sprays on GM herbicide-tolerant crops would cause relatively more or less harm to farmland biodiversity than their conventional counterparts. The government funded Farm-Scale Evaluations to obtain empirical evidence, whereby credible test methods depended upon involvement of a broader agro-environmental expertise, including nature conservation agencies. More generally, the UK government sought stronger empirical evidence to support official expert claims. Sceptical voices were encouraged to contribute to the UK policy debate, towards building a stronger consensus for agricultural biotechnology. In practice 'this effort to broaden politics led to a more extensive unpacking of scientific unknowns', argues Jasanoff (2005: 277).

By the late 1990s public protest was deterring approvals and a market for GM products. European supermarket chains decided to find alternative sources of grain, as a means to exclude GM ingredients from their own-brand products (Levidow and Bijman, 2002). Governments had greater difficulty to justify why they were continuing to support the commercial use of GM crops and foods. Some member states banned GM products that had already gained EU-level approval. Amid a legitimacy crisis and trade conflict, civil society organisations found greater opportunities to block agbiotech and/or to demand more stringent regulatory criteria. In such ways, they used the trade conflict to intensify domestic political conflict (cf. Bernauer, 2003).

In June 1999 some Ministers in the EU's Environment Council agreed to block the regulatory procedure for GM products. Many of them signed one of two similar statements that they would not consider additional GM products for approval until the EU had made significant changes to its regulatory framework in order to address various weaknesses. To justify this delay, they cited 'the need to restore public and market confidence', while leaving ambiguous the object of lost confidence, for example, GM products, EU regulatory procedures and so on.

Together those statements became known as an unofficial *de facto* moratorium on any further approvals of GM products. In their June 1999 statements, EU Environment Council members specified the regulatory changes necessary before the approvals procedure could resume. The list included: basing risk assessment upon precaution, and requiring traceability and labelling of all GM products and derived products. The European Commission still had the legal authority to approve products if the European Council failed to do so, but the Commission had a weaker will and political authority after the 1997 crisis over Ciba-Geigy's Bt maize.

Stimulated by the *de facto* moratorium in 1999, the Commission initiated legislative changes to accommodate demands from member states. Until then, draft revisions of the Directive had aimed to streamline the approval procedure, potentially reducing regulatory burdens. After 1999 the redrafts incorporated more precautionary and stringent criteria (Levidow and Carr, 2000). For example, no risk should be ignored on grounds that it would be unlikely, and uncertainties should be explained for any identified risk (EC, 2001). Such rules formalised pressures to open up expert judgements about scientific uncertainty.

A common rationale for the *de facto* moratorium, the aim to 'restore public and market confidence', provided a means to go beyond previous disagreements about the 'scientific' basis for regulatory requirements. Institutional reforms provided a more flexible basis for regulatory-expert procedures to accommodate many concerns of mainstream consumer NGOs and environmental conservation groups – though not demands to prove 'harmlessness', which would have the effect of simply blocking GM products. These changes accommodated demands from many civil society organisations, while potentially creating a more legitimate basis to approve some GM products in the future.

Precautionary shifts in agbiotech regulation

Since the late 1990s conflicts continued over exactly how to revise and implement EU regulations. Meanwhile the US intensified its threats to bring a WTO case against the EU regarding agbiotech products. The Commission privately warned US officials that their overt threats were undermining its own efforts to establish a workable regulatory system.

Indeed, agbiotech critics used the transatlantic conflict to press for more stringent EU rules. The Commission generally favoured less-restrictive criteria than some member states did, especially on grounds that EU rules must be workable and comply with international commitments. Yet critics denounced its specific proposals as a surrender to US pressures, while advocating more stringent rules as necessary for EU sovereignty. When the European Parliament supported such rules on GM labelling, a Green MEP declared, 'It's a great victory for consumer choice and a clear message to Tony Blair and his American friends' (Agence Europe, 2002). After the US decision to launch a WTO case against the EU in May 2003, agbiotech critics likewise framed any EU approval of a GM product as a surrender.

In response the Commission has sought to counter public suspicions about external pressures. Eventually it promoted the new EU procedures as a better global model, whose credibility would depend upon timely implementation. In particular, member states should restart the procedure for approving GM products: 'We have to start because we want to demonstrate to the rest of the world that our way of taking decisions about GMOs works. Otherwise they will not believe us', according to the DG-Environment Commissioner (Margot Wallström, Associated Press, 28.01.04). After much effort along those lines, the Commission finally approved the first new GM product in spring 2004, the implementation date for new EU rules on traceability and labelling. Not coincidentally, this decision came just before a crucial meeting of parties to the WTO dispute.

Commission policy statements awkwardly combine diverse aims. For example: 'science-based regulatory oversight' aims 'to enable Community business to exploit the potential of biotechnology while taking account of the precautionary principle and addressing ethical and social concerns' (CEC, 2003: 6, 17). Shifts in practical meaning are summarised in Table 1: regulatory criteria in the left-hand side are sketched in this section, and expert advisory roles on the right-hand side are sketched in the subsequent section.

Crop cultivation: agro-environmental risks

When GM crops were being evaluated for cultivation uses in the mid-1990s under the Deliberate Release Directive, sharp disagreements arose even before public debate became widespread. Proponents narrowly defined the 'adverse effects' to be evaluated, thus accepting the normal hazards of intensive monoculture. Environmental NGOs had warned that such products would result in a 'genetic treadmill', by analogy to the familiar 'pesticide treadmill'. Safety claims accepted such effects, for example, the prospect of spreading herbicide-resistant weeds or insecticide-resistant insects; thus regulatory procedures accepted the inherent hazards of intensive monoculture.

This approach conceptually homogenised the European environment as a resource for efficient agri-production, as in the US model of intensive monoculture. Product approval decisions minimised responsibility for any undesirable effects. This approach facilitated EU-wide regulatory harmonisation, transatlantic trade in GM products and thus the TABD-TEP agenda.

When the Directive was revised along more stringent lines, the revision broadened the range of agro-environmental effects which are to be prevented or managed; these included the effects of any changes in management practices, for example, herbicide sprays. Market-stage monitoring could be required to verify any assumptions in the risk assessment (EC, 2001). Overall these changes broadened the scope of 'scientific' issues and of the European 'environment', thus complicating the earlier basis for regulatory harmonisation.

Another crisis provided an opportunity for different policy agendas. In the 2000 StarLink scandal, a GM maize had been legally cultivated by US farmers before government approval for food and feed purposes; such approval was ultimately denied because of doubts about health risks. StarLink maize was then found to be illegally present in the food chain, especially in North America but also in Europe. Anti-agbiotech groups cited this scandal to demand a ban on all GM products.

Yet the Commission used the scandal for a different agenda. Under the slogan, 'one door, one key', the Commission had been attempting to integrate approval for all product uses within 'vertical' product-based legislation since the mid-1990s. This proposal now gained support as a means to avoid another StarLink scandal, by ensuring that no GM crop cultivation would be authorised before approval for food and feed purposes.

In using this opportunity for legislative change, the Commission centralised the regulatory procedure. Under a new GM Food and Feed Regulation, advice from the European Food Safety Authority (see later section) would inform decisions by the Commission, coordinated and prepared by DG-Agriculture. Thus decisions to approve a GM crop for cultivation were removed from the Deliberate Release Directive, thus potentially bypassing the national authorities responsible for environmental protection. This change was widely criticised for marginalising environmental issues but prevailed in the final version (EC, 2003a). Parliament supported centralisation as a means to avoid the disagreements which had delayed regulatory decisions since the late 1990s.

GM food risks

When the EU was approving the first GM crops for cultivation, their food uses came under a new regime which facilitated approval decisions. The 1997 Novel Food Regulation had a simplified procedure for approving novel products including GM foods. A national authority need not carry out a risk assessment for any GM product which had 'substantial equivalence' with a non-GM counterpart regarded as safe (EC, 1997a). In practice this procedure assumed that physico-chemical composition tests alone could demonstrate such equivalence. Under a similar policy, the US FDA did not generally require a risk assessment or even approval of GM foods. Complementing the TABD-TEP agenda, the EU-simplified procedure facilitated regulatory harmonisation, thus helping to avoid trade barriers within the EU and across the Atlantic.

However, the concept of substantial equivalence underwent widespread criticism, especially as an 'unscientific' means to bypass risk assessment and safety tests. Some national expert advisory bodies were already interpreting the concept according to more stringent criteria; for example, they requested more rigorous evidence of physico-chemical composition, as well as more toxicological tests. Scientists' efforts along those lines converged with demands of consumer organisations criticising substantial equivalence (Levidow, Murphy and Carr, 2007).

Eventually substantial equivalence lost its statutory role. When Italy banned GM foods which had been approved by the US and EU, other member states joined its attack on the simplified procedure. Recognising its legitimacy problem, the Commission abandoned the simplified procedure when drafting the GM Food & Feed Regulation in 2001. Substantial equivalence would be kept as a risk-assessment tool, but EU-wide harmonisation might be difficult to achieve for this 'dynamic concept', whose interpretation was still under development, according to a Commission official (Pettauer, 2002: 23). The concept of substantial equivalence continued to inform expert judgements

Table 1 Policy changes in EU agbiotech regulation

	Agbiotech risk legislation	GM food regulation	GM labelling and consumer choice	Scientific uncertainty: policy role	EU advisory expertise: general arrangements	EU expert advice on GM products
Mid-1990s	DRD bases the internal market on high level of environmental and health protection; harmonise criteria	Substantial equivalence can justify GM food safety	No GM labelling should be required for safe products	Safety based on 'sound science'; no evidence of risk (uncertainty ignored)	Hosted by the Directorate-General responsible for legislation and product approvals	[No EU-level expert advice on GM products]
Post-BSE crisis (1996–97)	Proposals to streamline DRD procedures and to lighten regulatory burdens	NFR bases the simplified procedure on substantial equivalence	GM food must be labelled according to scientific criteria of detectability	Debate over how expert procedures should address uncertainty about risks	Establish advice independent of policy influence, material interests & member states	'No evidence of risk' from each GM product; 'adverse effects' are defined narrowly
Late 1990s	Broaden risk assessment and require market-stage monitoring (2001 revision of DRD)	Some member states imposed more stringent criteria for 'substantial equivalence'	GM labelling rules are extended to additives	Guidelines for triggering the Precautionary Principle, i.e. measures to manage uncertain risks	Functionally separate risk assessment from risk management, thus protecting the scientific integrity of expert advice	No reason to indicate that (each) GM product will cause adverse effects

Since 2001	Centralise expert advice to facilitate harmonisation (GM F&F Regn)	Abandon the simplified procedure for GM food (GM F&F Regn)	Label food according to any GM source and ensure traceability (T&L Regn)	Any restriction must be justified by a risk assessment indicating uncertain risks	EFSA to clarify different views of expert bodies, and to build expert networks (2002 food law)	Safety claims undergo pressures to acknowledge uncertainties and normative judgements
Explicit policy aims	Exploit the potential of biotechnology while taking account of the precautionary principle and social concerns	Substantial equivalence remains a 'dynamic concept' for assessing GM food safety	Free choice to buy GM or non-GM food; ensure that the free market can function	PP must be used in ways compatible with EU treaty obligations	Independent, objective advice should inform decisions which can gain public confidence	Obtain adequate data to clarify uncertainties and overcome expert disagreements in risk assessment.

Abbreviations

DRD = Deliberate Release Directive (EEC, 1990; EC, 2001)
EFSA = European Food Safety Authority (EC, 2002)
GM F&F = GM Food & Feed Regulation (EC, 2003a)
NFR = Novel Food Regulation (EC, 1997a)
PP = Precautionary Principle (for example, CEC, 2000)
T&L = Traceability and Labelling Regulation (EC, 2003b)

in more stringent ways, as regards what evidence would be adequate for a risk assessment.

GM labelling and traceability

GM labelling rules have been introduced and extended to manage market instabilities. EU policy initially rejected demands that GM food should have a mandatory label. According to Commission officials, such a rule would stigmatise GM products as abnormal, threaten the internal market, and undermine science-based regulation. In response to public protest and consumer concerns, in 1997 the retail trade imposed its own labelling rules; these effectively defined what is/is not a GM food, though the criteria varied across EU member states. Combined with public unease, diverse labelling criteria could have jeopardised the overall market for soya or maize, as well as the EU internal market. Recognising this problem, in 1997 new EU rules required labelling according to the detectability of DNA or protein at a 1 percent level.

Further changes in labelling rules were linked with market functions. As an extra condition for lifting the 1999 *de facto* moratorium, EU member states had demanded full labelling and traceability of GM ingredients, that is, regardless of their detectability. In response, in 2001 the Commission issued draft legislation along those lines. Some Commissioners had previously opposed such rules but now supported them as essential 'for the free market to function effectively'. The new regime would allow a low threshold for 'adventitious presence' of GM material, that is, levels which were technically unavoidable (EC, 2003a). Each GM crop must be traceable throughout the agro-food chain, for example, by using 'unique identifiers' for the specific transformation event which constructed a GM crop (EC, 2003b). Retailers became dependent upon a paper trail to verify sources of grain, especially in cases where it is highly processed and so makes the DNA undetectable.

The new rules implicitly linked precaution with markets. Authorities could now trace and withdraw a product if problems arose later. Consumers would have a free, informed choice to avoid GM products and thus to make their own judgements on safety. Not simply 'completing the internal market', EU rules were redefining product identity, thus restructuring markets for GM ingredients as well as non-GM food. In the original 1990 regulatory framework, the public had been cast as an audience for safety claims, as consumers of food potentially containing GM ingredients, and thus as supporters of a beneficial technology – roles which publics eventually rejected. New rules accommodated demands to extend consumer rights and to clarify the identity of any food which may contain GM ingredients. Using the new opportunity, anti-agbiotech activists sought to block GM grain whose products would now require a GM label under the new rules.

Expert authority for regulatory decisions

In parallel with the legislative changes sketched above, the EU also made changes in expert advisory arrangements (see Table 1, right-hand side). This meant more accountable ways of translating risk controversy and scientific uncertainty into criteria for evidence. These changes institutionalised aspects of the 'consumer rights' agenda, while incorporating consumer organisations into consultation procedures.

Precaution and uncertainty

In the late 1990s precaution was becoming more contentious as grounds to block products, especially in the agri-food sector. After the US brought its WTO case against the EU over hormone-treated beef, eventually the Appellate Body ruled that the defendant had failed to justify its beef ban through a risk assessment (WTO AB, 1998). Amid EU conflicts over the Precautionary Principle in many sectors, the Commission sought to clarify the concept.

According to its 2000 Communication, risk-management measures could be justified where uncertain risks jeopardise 'the chosen level of protection', that is, the type or extent of risks deemed acceptable. In such cases, a risk assessment must demonstrate reasonable grounds to suspect that a product could cause 'potentially dangerous effects'. Whenever taking precautionary measures, authorities must make efforts to obtain additional scientific information for 'a more complete risk assessment' (CEC, 2000).

The guidelines aimed to ensure that the Precautionary Principle would be used only in ways defensible under the EU's treaty obligations. The Commission's criteria were broader than in WTO rules and in some national policy frameworks, though more narrow than in some others. In effect, Commission guidelines provide a basis for selectively limiting the Precautionary Principle, that is, to be invoked only in cases where the Commission decides to justify trade barriers. Such decisions would also depend upon expert bodies giving explicit advice about scientific uncertainty.

Advisory expertise

In the run-up to the BSE crisis, expert advice downplayed scientific uncertainties and made policy judgements about their manageability, yet safety claims were officially portrayed as 'science'. Moreover, experts aimed to give advice that would be politically acceptable to regulators, while avoiding public alarm about any risks (Millstone and van Zwanenberg, 2001). The Commission likewise covered up the BSE problem, for fear that public concern about the BSE problem would endanger the European beef market, according to a report by the European Parliament (1997). Expert bodies were being used by politicians to avoid full responsibility for decisions. Consequently, EU safety claims came under greater suspicion as policy stances, especially given that expert advisory bodies were hosted by the same Directorate-General responsible for relevant legislation and product approvals.

In response to various food crises including BSE, policymakers aimed to make EU scientific committees more independent in three respects: from DGs which propose and implement legislation, from member states, and from material interests. In reorganising its scientific committees, the Commission aimed 'to obtain timely and sound advice', 'based on the principles of excellence, independence and transparency' (EC, 1997b). Prospective members nominated themselves, rather than being chosen by member states. Risk assessment was separated from risk management through new arrangements, later described as a 'functional separation'. Scientific committees were shifted to the Directorate-General for Consumer Health, renamed DG-SANCO.

Greater political dependence upon expert advice meant more conflicting advice and difficulties for regulatory harmonisation. Some member states created independent agencies whose risk-assessment advice often questioned safety assumptions, even the expert advice from EU-level scientific committees. In this way, 'the legitimacy and the autonomy of the European Commission, and indeed its *rapports de force* with the EU member states, are thus being displaced to the arena of scientific expertise' (Dratwa, 2004: 13).

EU expert advice was further restructured, as a means for the Commission to enhance its political authority, while incorporating potential critics into new institutions. Under a 2002 food law, the European Food Safety Authority (EFSA) was created as an independent body which would help set 'science-based' standards for risk assessment. It would have greater expert resources, needed to clarify or reconcile disagreements across different expert bodies (EC, 2002). In establishing EFSA, its Management Board included representatives of industry and consumer NGOs as partners with shared understandings of policy problems, especially the need to gain public confidence (Smith, Marsden and Flynn, 2004).

EU expert advice on GM products

In the EU regulatory procedure for evaluating GM products, there was much conflict over the criteria in the mid- to late 1990s. According to the advice of EU scientific committees on each product, there was no evidence to indicate that the product would cause adverse effects. Such a wording left ambiguous the burden of evidence to demonstrate safety. Safety claims often depended upon normative judgements, for example, by classifying some undesirable effects as merely agronomic and therefore irrelevant, or by advising on the management of such effects, or by favourably comparing any harm from a GM crop to the harm from agrochemical usage. All these judgements came under criticism from member states (Levidow, Carr and Wield, 2000).

EFSA was established in 2003, when member states were again evaluating GM products under the revised Deliberate Release Directive. In evaluating a series of GM products, for example, food or grain imports for feed, EFSA's Scientific Panel declared that each one would be as safe as its non-GM counterpart. Some safety claims still depended on normative judgements about acceptable effects.

Partly for those reasons, member states have often disagreed with EFSA's safety claims. Some have questioned whether the available information was adequate for a risk assessment, whether the environmental assessment adequately covered their specific conditions, and whether specific undesirable effects should be regarded as acceptable. Such criticisms were often taken up by other member states. These expert disagreements can be interpreted as different accounts of precaution, whereby safety claims correspond to more narrow accounts than objections; the latter define harm and uncertainty in broader ways (Levidow, Carr and Wield, 2005). In response to demands from some national regulators and NGOs, eventually the guidance notes asked all applicants for a 'risk characterisation', which would make scientific uncertainties more explicit (EFSA GMO Panel, 2004). In sum, expert advisory arrangements institutionalised aspects of the 'consumer rights' agenda, for example, greater precaution and transparency, while incorporating consumer organisations into consultation procedures; these changes provided a stronger basis to legitimise safety claims and regulatory decisions.

Conclusion: contending policy agendas

Let us return to the questions posed in the Introduction: How did the US–EU conflict arise from contending agendas which span the Atlantic? How has the trade conflict been framed as a problem and used as an opportunity within the EU? Through these questions, this Conclusion discusses other major academic accounts of policy conflict over agbiotech.

As this essay has shown, the US–EU conflict arose from contending transatlantic agendas, which operated within and across the two jurisdictions. The policy problem was framed in three different ways, corresponding to the three Transatlantic Dialogues: regulatory harmonisation for trade liberalisation of a benign technology (TABD-TEP); the consumer's right to know and choose safe food, based on precautionary risk assessment, with sovereignty to accommodate such criteria (TACD); and opposition to products lacking proof of safety and imposing corporate control over the agri-food chain (TAED). Thus a neoliberal policy agenda set the context in which European protest could frame agbiotech as a dual threat of 'globalization' and unknown risks.

Jurisdictional characteristics and/or interactions?

Some academic analyses have attributed the US–EU conflict to separate jurisdictional characteristics. For example, US–EU differences in scientific risk assessment 'are related to different cultural attitudes' towards agbiotech, argues

Toke (2004). According to Isaac (2002: 251), 'endogenous political-economy factors' shape regulatory regimes in each jurisdiction, for example, in the USA and EU. He emphasises their internal sources, 'because the domestic regulatory approach sets the prospects of and limits to regulatory integration' at the international level, for example, through treaties (ibid: 24–5).

Of course these jurisdictional characteristics matter, but not as independent variables; at most they can explain internal conflicts and the relative strengths of contending transatlantic agendas within each jurisdiction. As EU domestic controversy and regulatory blockages led to a transatlantic trade conflict, policy actors diagnosed this as a problem according to their own agendas. The anti-agbiotech environmentalist agenda gained much greater support in Europe (than in the USA), for example, by linking GM food with public suspicion towards hazards of intensive agriculture and undemocratic 'globalization'. Amid a legitimacy crisis for government, civil society organisations found greater opportunities to block agbiotech and/or to demand more stringent regulatory criteria. In such ways, they used the trade conflict to intensify domestic political conflict (cf. Bernauer, 2003). The legitimacy problems and processes of European integration provided greater opportunity for those advocating more stringent rules or regulatory delays.

From the late 1990s onwards, some policymakers and expert advisors articulated a new problem – how 'to restore public and market confidence' ? as an imperative for institutional reform. Eventually the EU enacted more precautionary legislation. It also established EFSA as a means to enhance the public credibility of expert advice, partly by incorporating consumer organisations into new structures. As regards market confidence, new labelling rules redefined 'GM food' in successively broader ways. Taking advantage of these more comprehensive rules, agbiotech opponents sought to extend the commercial blockage of GM grain.

The above story differs from some academic diagnoses of regulatory frameworks as a problem. According to Toke (2004), national policies claim a scientific basis which in turn reinforces a given policy. Distinctive regulatory discourses are translated by scientists or expert advisors into 'factual' terms. A government may need to accommodate public concerns which eventually arise, so it would be better to anticipate them in advance. Yet an institutional path dependency often makes a policy framework insensitive or inflexible to such changes in context, he argues (ibid: 205–8). In his diagnosis, 'science-based regulation' readily becomes a constraint on policymaking. On the contrary, however, European expert advice flexibly changed its judgements along more stringent lines, thus stimulating and accommodating changes in regulatory policy, in response to public protest.

Going beyond an EU–US comparative approach, Bernauer (2003) analyses how transatlantic interactions underlie US–EU regulatory polarisation. Policy agendas have been driven by interest groups seeking political and market influence, he argues. The trade conflict amplified domestic controversies over agbiotech. Civil society groups have used the trade conflict to press for higher regulatory standards, though more successfully in the EU than in the USA. Thus a trade liberalisation agenda stimulated a 'trading up' process, as theorised more generally by Vogel (1995).

Perhaps unlike Vogel, and certainly unlike Toke, however, Bernauer diagnoses EU regulatory changes as a policy problem: evermore complex, stringent, costly regulations are insufficiently backed by robust institutional structures for implementing them; overall the trade conflict may reduce investment in agri-biotechnology. As a remedy, more centralised forms of governance in food safety would increase consumer confidence, he argues (Bernauer, 2003). Yet all those changes and difficulties are inseparable elements of the overall story in this chapter: more stringent regulations remain crucial for accommodating the consumer rights frame, integrating its representatives within new structures, and thus potentially legitimising a centralised procedure.

Economic-scientific versus social rationalities?

As a general framework for analysing regulatory-trade conflicts, Isaac argues that all regulations have two main functions. Their economic function is to improve the efficiency of the market system, while their social function is to ensure that market activity takes place in a way consistent with the preferences and expectations within a jurisdiction. These functions frame policy issues in different ways:

> [The] economic perspective generally assumes that technology and innovation are vital factors of economic growth and welfare. As a result, it supports a regulatory framework that encourages technological progress. For instance, it is quite common for economic analysis to support 'scientific-rationality' approaches to regulating the risk of new technology . . . The economic- and scientific-rationality perspectives are similar, in that they decompose complex behaviour and actions into causal-consequence models, which are then used to forecast outcomes (Isaac, 2002: 16–7).

> [The] 'social-rationality' approach holds that it is insufficient to view new technology and innovations simply as a positive force in economic growth. Instead, the social implications of science must be considered and, under this consideration, new technology may not always be greeted without reservation – despite its potential to improve economic growth (ibid: 21).

To develop a stable regulatory framework, a jurisdiction must find some way to 'balance' scientific and social rationality. In his view, 'the ideal regulatory framework essentially builds social credence into the scientific-rationality paradigm'. In particular, such a framework would acknowledge that normative issues are prior to empirical ones, so that science can help risk regulation to address societal concerns, argues Isaac (ibid: 257). While acknowledging normative issues, he diagnoses an imbalance or separation between social and scientific-economic rationality, thus implying that they could remain separate.

On the contrary, the EU story here illustrates how such rationalities were always linked, in changing ways. As an explicit means 'to restore public and market confidence', EU reforms changed the previous relation between social, scientific and economic criteria. 'Science-based regulation' now depended upon different social norms than before, for example, as regards the wider potential effects and uncertainties to be evaluated. Efficient 'free markets' now depended upon a clearer product identity through GM labelling rules, lest ambiguity undermine food markets. These changes can build social credence into scientific rationality, while expressing public preferences through social rationality (cf. Isaac, 2002).

In this case, any such rationality remains within the limits of a product-safety framework. EU advisory expertise was restructured as a means to strengthen safety claims, to enhance the Commission's political authority, to promote regulatory harmonisation and to keep any precautionary measures compatible with EU treaty obligations. When more transparent uncertainties and stringent criteria are cited to justify approvals of GM products, such decisions can accommodate the 'consumer rights' frame ? separated from the wider issues of the anti-agbiotech frame and demands for alternative agricultures. In this way, recent institutional reforms may help politicians to justify EU regulatory approval decisions which avoid trade barriers, thus facilitating trade liberalisation.

However, legitimacy problems continue as the EU relegates societal decisions to product safety and markets. According to Commission policy, risk regulation 'is the expression of societal choices': rules should ensure that market mechanisms function effectively, so that safe products are available to accommodate consumer preferences (CEC, 2002: 14). The EU policy problem can be diagnosed as a pervasive tension between the three contending agendas – especially between trade liberalisation versus democratic sovereignty over societal futures, beyond simply regulatory criteria for safe products.

What European integration?

What does this mean for the shape and legitimacy of European integration? Jasanoff analyses biotech in such terms. As biotech innovation devised novel 'designs on nature', its governance challenged some founding assumptions of liberal democracy, for example, 'that citizens have the capacity to participate meaningfully in decisions that seem increasingly to call for specialized knowledge and expertise'. Through various expert bodies, democratic control over biotech 'was sometimes set aside in favor of other culturally sanctioned notions about what makes the exercise of power legitimate' (Jasanoff, 2005: 272, 287). Within the EU agenda of regulatory harmonisation, agbiotech regulation undergoes a tension between two political models: seeking to eliminate national divergences in policy framings, versus protecting deep-seated national values that generate them (ibid: 71).

The EU story in this chapter suggests more complex, contradictory futures. Whenever one member state proposes more stringent criteria or broader framings for risk assessment, such proposals have often gained support from others, as a potential European standard (Levidow, Carr and Wield, 2005). Possible futures go beyond simply maintaining national differences within the EU or else harmonising them away. Consequently, regulatory harmonisation remains a dynamic process of disputing, broadening and levelling up standards. In its own way, EU agbiotech regulation shapes the future European Union and its basis for democratic legitimacy.

Note

1. This essay draws upon results of three research projects in which the author took part: 'Safety Regulation of Transgenic Crops: Completing the Internal Market?', funded by the European Commission, DG XII/E5, Ethical, Legal and Socio-Economic Aspects (ELSA), Biotechnology horizontal programme, during 1997–99. Reports available at http://technology.open.ac.uk/cts/srtc/index.htm. 'Precautionary Expertise for GM Crops (PEG)', funded by the European Commission, DG-Research, Quality of Life programme, during 2002–04. Reports available at http://technology.open.ac.uk/cts/peg/index.htm. 'Trading Up Environmental Standards? Transatlantic Governance of GM Crops', funded by the UK's Economic and Social Research Council between 2002–04. Brochure available at http://technology.open.ac.uk/cts/tup/index.htm; also book (Murphy and Levidow, 2006). The author would like to thank Robert Falkner and Celina Ramjoué for helpful editorial comments on this chapter.

8

Competition for Public Trust: Causes and Consequences of Extending the Transatlantic Biotech Conflict to Developing Countries

Thomas Bernauer and Philipp Aerni

Introduction

In this chapter, we argue that the agri-biotech controversy in affluent societies has become largely symbolic in content. Rather than trying to address the challenges and opportunities of agricultural biotechnology, political stakeholders are seeking to appropriate public trust, often as self-appointed representatives of the poor and the environment. In this context, public trust may be regarded as a political resource like money and political power. However, public trust, managed as a private good in politics, cannot be exchanged for money or political power and therefore fuels radicalism and prevents compromise.

This argument helps explain why the transatlantic dispute on genetically modified organisms (GMOs) is increasingly carried out in developing countries. The search to gain public trust and political legitimacy at home induces political stakeholders to shape the public debates on agricultural biotechnology in developing countries. Yet, in view of the growing domestic research capacity and the pragmatic public attitudes toward agricultural biotechnology in advanced developing countries, foreign interference is becoming increasingly ineffective in shaping the regulatory agenda of these countries.

Differences in public perceptions, interest group dynamics, political systems, and industrial structure have, over the past 15 years, driven European and US agri-biotech policy in opposing directions (Bernauer and Meins, 2003; Bernauer, 2003; Jasanoff, 2005; Winston, 2002). In the United States, biotech firms and large farmers have pushed for and obtained comparatively permissive regulatory standards. In the European Union, non-governmental organizations (NGOs) involved in professional advocacy work have left a strong imprint on the Union's highly precautionary regulation of GMOs in agriculture and food. In 2003, these regulatory differences escalated into a formal WTO trade dispute.

The implications of the transatlantic agri-biotech conflict for developing countries have been substantial. Both economic powers have been using carrots and sticks to motivate developing country governments to side with their respective position. Moreover, firms, NGOs, and governments from Europe and the US have been building alliances with local *non-state actors* in developing nations in an effort to generate political pressure for or against agri-biotechnology from within these countries.

The existing literature does not offer a systematic explanation of why the US and the EU would extend the transatlantic biotech conflict to developing countries (see, for example, Paarlberg, 2001, 2003; Victor and Runge, 2002; Falkner, 2002b, 2004a; Prakash and Kollmann, 2003). This behavior is particularly puzzling in the case of Africa whose commercial market for the technology and agricultural trade with the US and the EU is small. That is, variation in economic interests alone does not provide a convincing explanation of variation in US and EU behavior in this policy area. We seek to fill this gap by offering an explanation that focuses on competition for public trust and discursive power. Though we develop this theory with regard to a specific empirical area we are confident that it provides important insights into government, NGO, and corporate behavior in other areas of the global political economy as well.

European and US influence on developing countries

Developing countries have become an important issue in the global agri-biotechnology debate. This prominence of developing countries (Aerni, 2002a) is rather surprising given that these countries are not yet very important developers or importers of the technology.

It reflects at least in part a series of efforts by advanced industrialized countries to influence poorer countries' choices in this area. The most visible and controversial recent examples include

- US food aid to Africa and
- Egypt's non-support for US legal action in the WTO against EU agri-biotech regulation.

In 2002–03, the World Food Programme provided food aid to several sub-Saharan African countries that were at risk of suffering a severe famine. The contribution from the US contained transgenic Bt corn. Zambia rejected such 'potentially toxic' food and asked for food aid free of genetically modified (GM) food. US policy-makers reacted to this decision by blaming the EU for pressuring African countries to reject GM food even though the major international food safety agencies confirmed its safety for human consumption. The Europeans, in return, accused the US of trying to introduce GM crops in Africa through the backdoor of the World Food Programme.

In May 2003, the United States first declared that Egypt was supporting its legal action in the WTO against EU agri-biotech approval regulation. Some days later it began to backtrack on its statement as Egypt appeared to be wavering. In the end Egypt withdrew from the case, citing rather vague concerns about adequate and effective environmental and consumer protection – neither did it formally support legal action (unlike Argentina and Canada), nor did it request third-party status in the proceedings of the WTO's dispute settlement panel that was established in August 2003. Even though few details on Egypt's decision-making have become public it appears to have agonized considerably over the dilemma. Egypt is the second-largest recipient of US foreign aid. Moreover, in a letter to the foreign minister of Egypt in 19 June 2003, Senator Grassley, chairman of the US Senate's Committee on Finance, quite explicitly threatened to slow or halt negotiations on a free trade agreement with Egypt if the latter withdrew its support. On the other hand, Egypt is also heavily dependent on Europe. The EU accounts for around 40 percent of Egypt's exports and 34 percent of its imports. Egypt also receives substantial amounts of financial assistance from the EU, particularly in the framework of the Egypt–EU Association Agreement and the MEDA program (European Union, 2004).

Beyond the two episodes just discussed, influence on developing countries' agri-biotech policies has been exercised through a variety of mechanisms, including the following:

Over the past 5 years, the EU and the US have sought to influence the position of developing countries in the Cartagena Protocol on Biosafety (see Chapter 1), a protocol to the 1992 UN Convention on Biological Diversity (CBD), concluded in 2000. This protocol governs transboundary movements, handling, transit, and use of living modified organisms that may have adverse effects on biological diversity. It addresses primarily environmental effects of trade in GM products, but also takes into account public health aspects (see Article 1 in the Cartagena Protocol on Biosafety). The EU supports the Protocol. The US opposes it because the latter endorses the application of the precautionary principle in international trade with GM products. As of December 2004, 110 countries, including many developing countries, had ratified the Protocol[1] (Falkner, 2000, 2002b).

The Cartagena Protocol has established an important multilateral legal justification for EU assistance to developing countries in the biosafety area. Its capacity-building component includes scientific and technical training, help in establishing institutional and regulatory mechanisms for risk assessment and risk management, access to relevant information, and financial assistance for these purposes. These activities are supported by the United Nations Environment Programme and the Global Environment Facility.[2] Most of these effort have the effect of constraining rather than promoting agri-biotechnology applications.

The US, not being a party to the Protocol, has established bilateral networks of cooperation in agricultural biotechnology research and development. In other words, while the EU assists developing countries in the design of precautionary regulatory frameworks, the US has sought to encourage developing countries to adopt more permissive regulations that allow for agri-biotech R&D activities beyond the laboratory. For example, the USAID Biotechnology Initiative was launched in 2001 'to use the benefits of agricultural biotechnology throughout Africa to enhance food safety and security' (Kellerhals, 2001). Other initiatives include the Collaborative Agricultural Biotechnology Initiative (USAID, 2003a); the Collaborative Research Support Programs (CRSPs) (USAID, 2003b); the USAID-supported African Agricultural Technology Foundation (AATF); the Bean/Cowpea Collaborative Research Support Program; USDA technical assistance for the cotton-growing industry in West Africa; and the regional African Center of Excellence for Biotechnology.[3]

At the non-governmental level, hundreds of NGOs, business associations, and individual firms have been involved in efforts to influence developing countries' agri-biotech policies. The British anti-biotechnology website GMWatch[4] profiles organizations it defines as being committed to pro-biotech initiatives. Out of the 114 organizations on its list, 24 are either based in developing countries or are dedicated exclusively to agricultural biotechnology in developing countries. Most of the remaining institutions also have strong links to developing countries and regularly issue statements on agri-biotech issues in poorer parts of the world. The website NGO Watch,[5] in turn, which is managed by the US-based American Enterprise Institute (which is strongly pro-biotech), offers a list of 159 NGOs that are active in environmental, human rights, services, speech, family/women, and labor issues. Of these NGOs, the names of 103 contain the words World, International, or Earth or are known to be committed to developing country issues. Even though none of these NGOs deals exclusively with agri-biotech issues, most of them are engaged in multi-sectoral advocacy work that connects them to the controversy over agri-biotechnology. Such NGOs can thus be quickly mobilized for campaigns against agri-biotechnology in developing countries.

NGOs, business associations, and firms from advanced industrialized countries have influenced state and non-state actors in developing countries in a variety of ways (Paarlberg, 2001; Cohen and Paarlberg, 2004; Kremer and Zwane, 2005). Examples on the proponent side include corporate donations of technology to developing country research institutes, education/instruction of stakeholders from poor countries in advanced industrialized countries, funding for biotechnology and biosafety research. On the opponent side, activities include funding for protest campaigns, capacity-building activities, and organic agriculture initiatives (Bob, 2002; Paarlberg, 2001, 2003; Cohen and Paarlberg, 2004). Survey research by one of the authors (Aerni, 2001, 2002a) has shown, moreover, that stakeholders from advanced industrialized countries exert substantial influence on the most vocal participants in public debates on agricultural biotechnology in developing countries. A network analysis of stakeholders involved in the public biotech debate in the Philippines showed, for example, that domestic NGOs campaigning against agri-biotechnology were largely financed by foreign stakeholders.[6]

Puzzle

Most of the (scarce) political science literature on national and international agri-biotech policy focuses on explaining differences in regulation across jurisdictions (for example, Prakash and Kollmann, 2003; Bernauer and Meins, 2003), escalation of trade disputes or, conversely, the likelihood of international regime formation in this area (Raustila and Victor, 2004; Coleman and Gabler, 2002; Falkner, 2000, 2004a; Bernauer, 2003). Parts of the literature also concentrate on agri-biotech policy in developing countries and how to promote the technology in such countries (for example, Cohen and Paarberg, 2004; Victor and Runge, 2002; Paarlberg, 2001). However, the extant literature does not offer any systematic answers to the question of *why the transatlantic battleground has spread to poorer countries*.

Paarlberg (2003, the principal point of reference for this chapter, focuses on explaining why the diffusion of the EU's restrictive policy-approach in agri-biotechnology to the developing world has been more effective, compared to the US' more liberal policy-approach. He proposes four channels of influence in agri-biotech policy: (1) influence exercised via international institutions; (2) influence via development assistance; (3) influence exercised by NGOs, firms, and other non-governmental actors; and (4) influence by market forces (trade patterns, consumer tastes in rich countries) (Paarlberg, 2003; Angelo, Masiga and Musiita, 2003). While this categorization is quite compatible with the empirical observations discussed above it offers only very limited insights into why the transatlantic battleground has been extended to poorer countries in the first place. We thus view our research as complementary to the work by Paarlberg and others.

The potential driving forces of US and EU behavior in regard to developing countries appear in part as assumptions in Paarlberg's argument. The extension of the transatlantic conflict to poorer countries is, so it seems, viewed as an international struggle for markets and for influence on international regulatory processes. The US, in this perspective, is primarily pursuing a strategy of opening markets for its agricultural products and for agri-biotechnology. The EU, in turn, is trying to block such attempts and 'export' its own, more restrictive, regulatory approach. Both sides are trying to coerce and/or entice poorer countries into supporting their respective policy position in the WTO, the CBD, and other important international fora. To the extent they are able to win more allies in these international bodies their influence on international standard setting grows. Research on WTO dispute settlement suggests indeed that winning more allies may improve the chances of winning WTO dispute settlement cases. WTO dispute settlement bodies tend to behave as strategic actors that strike a balance between maintaining political support and maintaining legal impartiality and thus credibility. With an increasing number of allied countries in the WTO the likelihood of an adverse verdict decreases at least to some extent. Particularly in cases where the legal situation leaves much room for interpretation, which seems to be the case in agri-biotech issues, WTO panels and the Appellate Body are more likely to avoid rulings against politically important groups of countries (Kelemen, 2001; Garrett and McCall, 1999; Bernauer, 2003).

Explanations along these lines illuminate some elements of the transatlantic agri-biotech conflict and its extension to developing countries. But they appear incomplete in several respects.

US efforts to 'export' its regulatory approach to developing countries may at least in part reflect pure economic reasoning – that is, an interest in opening new markets for US GM technology and GM farm products. EU countries have neither an economically and politically powerful agri-biotech industry, nor do they grow and export GM crops in any significant amounts. Whereas the EU is a net importer of farm products (with a negative agricultural trade balance of around $6.5 billion in 2001) the US is a net exporter of farm products. Yet both EU and US agricultural exports to developing countries are very small.[7] In other words, variation in straightforward economic interests does not seem to explain convincingly variation in US and EU agri-biotech policy vis-à-vis developing countries.

These problems with explanations based on economic interests become most obvious if we look at Africa, in respect to which the EU and the US have recently traded the hardest blows in their agri-biotech dispute. In that case, economic explanations should lead to the conclusion that neither the US nor the EU has a great interest in pushing African agri-biotech policy in one direction or another. The EU consumes around 85 percent of Africa's agricultural exports (€8.3 billion in 2001, compared to €0.8 billion for the US). For that reason, the EU could simply wait for 'trading up' (Vogel, 1995) in African agri-biotech regulation. African exporters have very powerful market incentives to adjust to European preferences and do not need to be pushed by additional means. As for the US, its market is open to GM products, so it should not care whether African exports to the US include GM products or not. Neither the EU nor the US export significant amounts of agricultural goods to Africa. Moreover, the purchasing power and the agricultural productivity potential in Africa is much too low to motivate substantial R&D on the part of European and US agri-biotech firms to serve this market. *Why do the EU and the US fight so hard over the agri-biotech policies of developing countries, and African countries in particular?*

One of the reasons noted by some analysts for the US decision, in May 2003, to initiate legal action in the WTO provides the starting point for an answer (Bernauer, 2003). It connects US and EU agri-biotech policies vis-à-vis developing countries to domestic politics. US decision-makers appear to have been driven in part by fear of an uncontrollable spill-over process reminiscent of the 'domino effect' debated during the Cold War. They thought that the European regulatory model would, in the absence of countervailing US action, first be emulated by developing countries with strong trade ties to the EU and, from there, would spread to other countries as well, notably those with strong trade ties to the US. Eventually, as much of the world was moving toward stronger legal constraints on agri-biotechnology, domestic and international pressure for stricter regulation in the US would mount. That is, if most other countries imposed strong restrictions on agri-biotechnology, voters and consumers in the US would begin questioning the legitimacy of their domestic regulations. Trust in government and regulatory authorities would suffer as a result. This explanation also raises more questions than it answers. Why have the EU and the US been fighting so acrimoniously over Africa in particular? Why should we expect a strong spill-over effect emanating from Africa, given that exports and imports of agricultural products from/to Africa and its potential market for the technology are small? We submit that the following theoretical argument, centering on competition for public trust, helps in explaining the observed European and US behavior.

Competition for public trust and discursive power

Virtually all advanced democracies, and notably the EU, its member countries, and the US, are characterized by pluralist interest group politics and substantial influence of the mass media on political processes and outcomes (for example, Baron, 2002; Frey and Kirchgässner, 2002). That is, policies tend to be strongly shaped by 'intermediary' politics, in which interest groups influence political agendas and policy-makers' choices not just through behind the scenes lobbying (rent-seeking) but, increasingly, through the mobilization of public pressure via public attention-seeking activities (Caduff, 2005; Aerni, 2003). Under such conditions public trust is a valuable political asset, particularly for non-elected non-state actors, such as firms and NGOs.

Public trust can be defined in terms of the belief among political constituencies that a particular actor or group is acting in the public interest rather than self-interest. In agri-biotech policy it also refers to the belief that a particular actor or group is telling the truth about benefits and risks of the technology and its applications (for example, Eurobarometer, 2003).

Public trust provides interest groups with legitimacy in the public arena, which they otherwise may find hard to obtain because they are not elected in a formal democratic process. Such legitimacy, in turn, provides them with political influence, primarily in the form of discursive power: policy-makers depending on election or re-election can usually expect to attract more votes, or public support in general, if they side with those non-state actors who enjoy a high degree of trust among the electorate. Similarly, (non-elected) public officials (for example, decision-makers in regulatory agencies) can expect more political support if they side with those who enjoy high degrees of public trust. They are usually appointed by elected politicians and receive budgets from those.

Public trust and legitimacy are important sources of discursive power. Discursive power refers to the ability to influence norms, values, ideas, political agendas, the framing or definition of solutions to particular societal problems, and the political discourse (or non-discourse) on specific problems more broadly. This form of power is also referred to as the 'third face of power' or 'soft power' in parts of the political science literature. It differs from two other forms of power, namely instrumental and structural power. The latter two derive primarily from material sources, such as economic or military capabilities, whereas discursive power hinges much more on public trust and legitimacy (Fuchs, 2004; Koller, 1991; Parsons, 1967; Lukes, 1974; Galbraith, 1984; Levy and Newell, 2002; Milner, 1991; Nye, 1991; Helleiner, 2001; Prakash, 2002). Many authors have in fact argued that influence on policy-input, and notably power over norms, ideas, political agendas, and rule-making, has in many areas of policy-making become more consequential than conventional sources of power over policy-output.

The above notion of public trust ties in more closely with sociological notions of moral legitimacy as a source of authority and discursive power (for example, Fuchs, 2004) than with social-capital-related notions of the concept. As to the latter, most authors (for example, Hardin, 2002; Fukuyama, 2004; Putnam, 1995, 2002) view public trust quite broadly as the backbone of economically prosperous and stable democratic societies. Others see public trust as a determinant of public support for certain policies or technologies, or as an indicator for success or failure of public policies. Priest, Bonfadelli and Rusanen (2003), for example, show that 'trust gap' variables predict national levels of encouragement for several biotech applications. They argue that this points to 'an opinion formation climate in which audiences are actively choosing among competing claims. Differences between European and US reactions to biotechnology appear to be a result of different trust and especially 'trust gap' patterns, rather than differences in knowledge or education.' The extent to which a particular stakeholder enjoys public trust affects the extent to which this stakeholder's positions on biotechnology are supported by the public.

In contrast to most analyses of discursive power we are less interested in which actors prevail in terms of being able to project more power but in understanding the role that the quest for public trust and discursive power plays in agri-biotech policy and the geographic expansion of the controversy.

Moreover, our argument on public trust and discursive power leads to less optimistic views on political processes and stakeholder behavior than in some other analyses of this nature. Fuchs (2004) and many other authors assume that NGOs (or civil society) have taken the lead in moralizing many policy issues, and that business has then followed. This has, so they argue, led to many coalitions between NGOs and business, private–public partnerships, the greening of industry, corporate citizenship, and so on.

Our argument views the quest for public trust, moral legitimacy, and discursive power as more conflictual. In fact, we assume that interest groups compete for public trust and try to manage it like a private resource. This competitive process tends to breed political polarization and radicalism. The reason is that competition for public trust, if the latter is treated as a private political resource, is based on exclusion ('trust us, not them'). Moreover, public trust, once appropriated by particular interest groups is not fungible. If an interest group that enjoys a high degree of public trust is willing to make a political bargain with another interest group that enjoys a lower degree of public trust but more political or economic power, the former group runs the risk of losing public trust entirely. The public would namely perceive this interest group to be acting in its private rather than the public interest. In other words, trust cannot be exchanged for political power or money and political compromises become very hard to achieve (Aerni, 2003).

Implications for agri-biotech politics

In the 1970s and 1980s most analysts and stakeholders in Europe and the US expected that science would enjoy sufficient public trust to act as the arbiter in domestic disputes over the risks and benefits of agri-biotechnology. They also expected that science would help achieve consensus on the design of appropriate domestic regulation. Internationally, they thought that the Uruguay Round trade talks, which led to the WTO in 1995, would establish a legal framework that defined scientific evidence as the principle justification for imposing trade restrictions in the name of concerns over human health and environmental risks. Science would thus help in separating protectionist from legitimate concerns on agri-biotechnology and eliminating the former – this would foster technological progress and free trade at the same time.

These hopes have been in vain. The status of science as an impartial arbiter in the global agri-biotech controversy has been gradually undermined in domestic debates as well as in the global trading system. More generally, neither science nor government nor the private sector have been able to infuse public trust in new technologies to the extent they did during the Cold War (Bauer, Gaskell and Durant, 2002; Beck, 1999; Anderson, Levy and Morisson, 1991; Jasanoff, 2005). Rapid developments in agri-biotechnology in the 1990s combined with further global trade liberalization have, therefore, led to a public trust deficit. Non-governmental organizations claiming to speak for public interests against the powerful interests of science, business, and government have, ever since, tried to capitalize on this deficit.

The ensuing competition for public trust among NGOs, industry, scientists, and governments has led to a shift in the public agri-biotech debate from risks in a scientific–technical sense toward worldviews and values (Eurobarometer, 2003; Gaskell, 2004). This shift has been accelerated particularly by the campaigns of large, internationally active advocacy groups, such as Friends of the Earth and Greenpeace. The trademark of such campaigns has been simple and forceful communication of worldviews and values that reduce complex and ambiguous scientific evidence on risks and benefits of agri-biotechnology to clear-cut and globally applicable good versus bad portrayals of the technology as well as associated stakeholders and their motives.

The shift in the public debate from scientific risk and benefit assessment to broader moral claims has also led to a changing meaning of risk. Risk, a concept that originates in mathematics and natural sciences, has increasingly been used in agri-biotech politics to mean danger from future damage caused by political opponents or anyone who is thought to be responsible for causing uncertainty (Jasanoff, 1995; Douglas, 1992; Beck, 1999; Sunstein, 2002). As a consequence, political pressure has been directed not so much against taking risks, but against exposing others to risks. While some observers and stakeholders praise this development as a 'democratization' of risk others denounce it as irrational 'politicization' of risk.

The uncompromising behavior of key anti-biotech advocacy groups, motivated by the expected benefits of appropriating public trust, has led to changes in the strategies of agribusiness firms. For example, Monsanto, in a public pledge, states:

> We will listen carefully to diverse points of view and engage in thoughtful dialogue to broaden our understanding of issues in order to better address the needs and concerns of society and each other . . . We will ensure that information is available, accessible and understandable . . . We will share knowledge and technology to advance scientific understanding, to improve agriculture and the environment, to improve crops, and to help farmers in developing countries . . . We will deliver high-quality products that are beneficial to our customers and to the environment, with sound and innovative science, and thoughtful and effective stewardship . . . We will respect the religious, cultural and ethical concerns of people throughout the world. The safety of our employees, the communities where we operate, consumers and the environment will be our highest priority . . . We will create clarity of direction, roles and accountability, build strong relationships with our customers and external partners; make wise decisions; steward our company resources; and take responsibility for achieving agreed upon results . . . We will ensure diversity of people and thought; foster innovation, creativity and learning; practice inclusive teamwork; and reward and recognize our people.[8]

Instead of explaining profits by pointing at economic performance, many of these firms have begun to portray themselves in public relations campaigns as some sort of charities that are primarily helping to protect the environment and to create a better life for the world's poor (see also http://www.whybiotech.com). In addition, many firms have sought to improve their public image through corporate social responsibility programs (see *The Economist*, Survey, 22 January 2005). The key purpose of these strategies is to be perceived by the public as less self-interested and, therefore, more trustworthy. Public trust, in this context, cannot be appropriated by delivering value for money (notably, to shareholders or employees), but only by declaring to act in the public interest. It appears, however, that these strategies have in most cases not led to more public trust in agribusiness, but have rendered the debate even more acrimonious, personal, and moral as anti-biotech advocacy groups have sought to discredit the relevant firms and their corporate social responsibility motives.

Feeding the world's poor

As described above, stakeholders from Europe and the US have in recent years extended their battleground to developing countries. The competition for public trust and the associated shift of the debate from risks to worldviews, motives and moral claims help in explaining why.

Large transnational networks of pro- and anti-biotech interest groups have emerged since the early 1990s. These networks hardly communicate privately with each other. Their representatives tend to face each other mostly in the public arena where they try to win the hearts and minds of electorates and policy-makers by accusing the other side of being morally indifferent to the fate of the poor.

In principle, we should expect environmental and consumer NGOs to have been the first movers in pursuing strategies of moralizing agri-biotech issues. On average, NGOs have fewer resources than companies to invest in behind the doors lobbying, provide legislative subsidies in the form of expert information, or reward political decision-makers with campaign contributions or other material benefits. Also, unlike firms, NGOs cannot benefit from threatening policy-makers with relocation to other jurisdictions (West and Loomis, 1999). Hence we should expect NGOs to compensate for these comparative disadvantages in relation to corporate actors by moralizing agri-biotech issues in order to increase their moral legitimacy, appropriate more public trust, and thus increase their discursive power (Cashore, 2002).

Empirically, it is rather difficult to demonstrate that competition for public trust has been a major driving force in extending the transatlantic biotech controversy to developing countries, for intentions are harder to identify than actual behavior. Similar difficulties exist in regard to showing empirically whether the pro- or anti-biotech side was first to carry the 'feeding the poor' issue into the controversy.

What seems to be clear is that the uncompromising nature of positions in respect to risks and opportunities of agri-biotechnology in developing countries has made political compromise impossible (Gaskell, 2004). We also found that NGOs have engaged much earlier than industry and government stakeholders in depicting the technology in broader moral categories, whereas industry and government stakeholders have long focused on more differentiated arguments about risks and benefits.

The public debate on agri-biotechnology in developing countries, has emerged as a popular media issue in Europe over the past 6 years, when it has become increasingly clear that GM food in Europe is becoming a non-issue for NGOs, because there currently are not any on the market in the first place. In turn, many developing countries started to embrace the new technology because they noticed that it may help solving certain problems in agriculture and boost long-term economic growth. Along with the shift of the debate from potential health risks for consumers in Europe toward to potential impact of agricultural biotechnology on food security in developing countries, 'public trust' and 'moral' elements of the controversy have grown in importance in the public discussion.

The competition for the moral high ground in the agri-biotech debate reflects efforts by both sides to appropriate public trust by demonstrating to electorates and consumers that they act in the public interest, whereas the other side seeks private benefits at the expense of societal (public) welfare or nature. Thus far, anti-biotech groups have been more successful in Europe than in the US in this regard, in part because public confidence in regulatory authorities, business, and science in Europe has been lower to start with (Eurobarometer, 2003; Bernauer, 2003). In other words, superior discursive power of pro-biotech interests in the US has enabled these interests to effectively prevent a wider public debate (and controversy) over the risks and benefits of the technology. Conversely, superior discursive power of anti-biotech interests in Europe has led to a pervasive controversy.

Mismatch of radicalism and pragmatism

Much of the controversy over risks and benefits of agri-biotechnology in developing countries is based on hypothetical claims by supporters and opponents of the technology from advanced industrialized countries. How do the positions of these stakeholders, as described above, map onto the positions of stakeholders and consumers from developing countries.

Unfortunately, there is very little data that would allow for systematic comparison of public and stakeholder attitudes in developing and industrialized countries. As to public perceptions, a 2000–01 study by the FAO (based on surveys carried out by Environics International/GlobeScan) offers the largest publicly available data of this kind. Table 1 summarizes some key results.

Table 1   Public perceptions in developing and industrialized countries

(1) The benefits of biotechnology outweigh the risks, % respondents who agree	
≥ 60%	Indonesia, Cuba, China, Thailand, India, Dominican Republic, Columbia, United States, Venezuela, Philippines, Mexico
< 60% and > 40%	Panama, Peru, Brazil, Canada, Netherlands, Nigeria, Turkey, Chile, Uruguay, Kazakhstan, Argentina, Australia, South Africa, S. Korea, UK, Poland, Germany
≤ 40%	Russia, Spain, Italy, Japan, France, Greece
(2) Biotechnology will benefit people like me in the next five years, % respondents who agree	
≥ 60%	Thailand, China, Philippines, United States, Mexico, Canada, Brazil
< 60%	Turkey, Netherlands, Columbia, UK, Australia, Germany, Spain, France
(3) Modifying the genes of plants or animals is ethically and morally wrong, % respondents who agree	
≥ 60%	France, UK, Spain, Philippines, Mexico, Turkey, Canada, Brazil, Thailand, Netherlands
< 60%	Australia, Colombia, Germany, United States, China
(4) Would continue to buy a food product if they learned it contained genetically modified ingredients to have a higher nutritional value, % respondents who agree	
> 51%	China, India, Brazil, United States, Canada
≤ 51%	Japan, Mexico, Germany, Australia, UK

The first question ('The benefits of using biotechnology to create genetically modified food crops that do not require chemical pesticides and herbicides are greater than the risks') was administered to 35,000 people in 34 countries (approximately 1000 per country). Countries in the three categories are ordered according to their support for GM crops. The second and third questions were administered only to 15 countries (none from Africa) and people who had heard of biotechnology. The questions were as listed in the figure. The fourth question is based on a survey of 10,000 people in 10 countries. Note that support for GM food is, on average, rather high because the questions are worded to emphasize the benefits of GM food and the technology. However, for our purposes relative levels of support are more important than absolute levels.
Source: Data from FAO (2004a) The State of Food and Agriculture 2003–04. Agricultural Biotechnology, Meeting the Needs of the Poor? (Rome: Food and Agricultural Organization) pp. 77–85.

The data for item (1) – if examined in more disaggregated form than shown – indicates that people from Europe tend to view agricultural biotechnology more negatively than people from the Americas, Asia, and Oceania. Perceptions in poorer countries are more positive than in richer countries (with the notably exception of rich countries where GM crops are grown): three fifth of respondents in non-OECD countries agree with the statement in item (1), only two fifth in OECD countries do so. As to the second item, three fourth of the respondents from non-OECD countries expressed positive views, around half of the respondents from OECD countries did so. Similar results were obtained for item (3). Results for item (4) show that more than 75 percent of respondents in China and India and 66 percent in Brazil were willing to buy GM food. In contrast, the OECD average was around 50 percent.

The sparse data on stakeholder attitudes in developing countries supports the above findings. Evidence from stakeholder surveys is important in this context for two reasons: first, it responds most directly to the question of a mismatch between stakeholder positions as put forth in the theoretical argument; second, contingent on the particular political system, stakeholder attitudes may, in many developing countries, have a greater impact on government policy than public perceptions as such. One of the authors conducted surveys of stakeholder perceptions and political influence in national public debates on agri-biotechnology in the Philippines, Mexico, and South Africa (Aerni, 2002a, 2002b). These surveys show that the majority of participants[9] in all three countries hold rather differentiated and pragmatic views on the risks and benefits of genetic engineering in agriculture, depending on the type of crop and key problems in domestic agriculture. Figure 1 illustrates this finding; it shows average stakeholder ratings on a scale of 1–5 for important problems in domestic agriculture and the potential of genetic engineering for solving these problems.

It turns out that drought (*Drought*) is considered to be one of the most important problems in agriculture in the three countries and genetic engineering is seen as having a potential for solving this problem. Further potential of genetic engineering is associated with important problems such as pest infestation (*Pests*), plant disease (*Disease*), and high use of pesticides (*Pesticides*). Market and infrastructure problems, such as inefficient market systems (*Market*), lack of investment in R&D (*R&D*), lack of irrigation facilities (*Irrigation*), poor transport networks (*Transport*), and post-harvest losses (*Postharv*) are a particular concern in the Philippines and Mexico; but genetic engineering is not regarded as having the potential for solving these problems (except for a reduction of post-harvest losses in Mexico through a transgenic tomato with increased shelf life).

The main view common to most stakeholders involved in the three surveys also was that Europe and the US should assist researchers in developing countries in learning how to use agri-biotechnology to address urgent problems in their respective countries.

Further evidence comes from actual agri-biotech R&D and GM crop cultivation. Several developing countries, notably Colombia, Cuba, Brazil, South Africa, China, and India have invested in agri-biotechnology for several years already. In terms of R&D spending, China, followed by India, are the leading biotech countries in the developing world.[10] Most of the worldwide growth of GM crop acreage in recent years has occurred in

152   *The International Politics of GM Food*

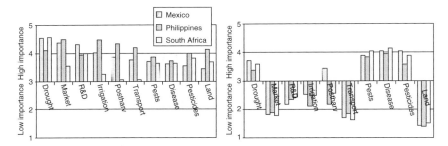

Figure 1   Perceived potential of genetic engineering for solving agricultural problems
Source: Aerni, P. (2002a) 'Stakeholder Attitudes towards the Risks and Benefits of Agricultural Biotechnology in Developing Countries: A Comparison between Mexico and the Philippines', *Risk Analysis* 22(6): 1123–37; Aerni, P. (2002b) *Public Attitudes towards Agricultural Biotechnology in South Africa*. STI/CID Research Report (Cambridge: Center for International Development, Harvard University). Available at http:// www. iaw.agrl.ethz.ch/~aernip/PDF/SAreport.pdf.

developing countries, notably China, Argentina, Brazil, and India. Some of these countries have, in their own research facilities, already developed a range of transgenic crop varieties and are eager to address problems in domestic agriculture with the new tools of biotechnology. In these countries, agri-biotechnology is on the verge of being perceived by policy-makers and electorates no more as an imported US technology but as a home-grown technology that is associated with national scientific reputation and pride.

In stark contrast to these national research efforts in some key developing countries, public investment in international agricultural research has remained very low (for example, Cohen and Paarlberg, 2004). For example, the Consultative Group of International Agricultural Research (CGIAR), which used to be the driving force of the first Green Revolution in the 1970s, has experienced drastic cuts in funding from Europe and North America (Aerni, 2004).

However, the continuing fall of prices for agri-biotech toolkits may soon lead to more rapid diffusion of the technology and make it affordable to many poorly equipped universities in developing countries. This would strengthen domestic agri-biotech research capacities and eventually enable developing countries to use the technology to solve their own particular local problems in agriculture. This trend would be accelerated if combined with an ambitious global open source effort in agricultural biotechnology similar to the Human Genome Project and the Institute for OneWorld Health in health biotechnology (*Nature Biotechnology*, 2005). If this scenario prevailed, an important argument of biotech opponents from rich and poor countries would collapse – that is, the argument that the technology is primarily an instrument to make large multinational companies from OECD countries richer and subject poor countries to their control.

Conclusion

In the short to medium term, the transatlantic biotech conflict and the associated competition for public trust as a political resource are likely to breed uncompromising behavior of stakeholders from rich countries vis-à-vis developing countries. Both sides will probably continue to use the mechanisms discussed above to try and pull individual developing countries into their respective camp. To varying degrees, poorer countries, and smaller developing countries in particular, will thus continue to pursue agricultural R&D strategies and regulatory models for agri-biotechnology that are largely imposed on them by advanced industrialized nations.

In the long run, however, pragmatism may prevail. To the extent that bottom-up demand for agri-biotechnology in developing countries grows and successful indigenous applications of the technology emerge, it will become more difficult for anti-biotech interest groups from rich countries to maintain current arguments and positions. To the extent that non-transgenic applications of the technology gain in importance and upstream patenting in the biotechnology business becomes harder, it will be much more difficult for stakeholders from rich countries to recruit support for extreme positions from stakeholders in poor countries. As a result, extreme agri-biotech proponents and opponents in Europe and the US would lose credibility and, therefore, public trust and discursive power. This, in turn, may eventually reduce global political tension over the technology and facilitate a more constructive approach to regulatory policy.

Political responses to and conflicts over new technologies should not be primarily seen in terms of economic interests. We concede that such explanations are important, but note that they fall short of providing valid accounts of an important phenomenon in the international political economy, and the nexus between domestic regulatory policy and global trade policy in particular. We are confident that the explanation focusing on the private management of public trust and discursive power offered in this chapter provides a basis for resolving empirical puzzles in other areas, and notably in international science and technology policy and associated environmental, health, and safety regulation. Global regulatory conflicts over the use of growth hormones in meat production, food irradiation, toxic waste treatment and disposal, hazardous chemicals, environmental and labor standards in the WTO, and nanotechnology are examples.

Although this chapter outlines the puzzle to be solved, provides a theoretical argument, and offers some preliminary empirical information that suggests the argument is plausible, full empirical testing of the claims would require a lot of new data. In particular, large-scale surveys of pro- and anti-biotech stakeholders (for example, NGOs, firms, business associations, governments) and consumers from a large set of developing and industrialized countries. These surveys would have to show to what extent agri-biotech positions are shaped by different levels of economic development, organized political activism, and other factors. They would also have to show whether the more radical pro- and anti-biotech groups are indeed driven by processes of competition for public trust.

Notes

1. It is too early to tell, however, whether this implies that the EU position on agri-biotechnology is winning over the US position – much will depend on how restrictive the Protocol's provisions will be implemented in developing countries.
2. http://www.biodiv.org/welcome.aspx; www.unep.ch/biosafety/.
3. http://www.usda.gov/Newsroom/0398.04.html.
4. www.GMwatch.org. Accessed on 10 December 2004.
5. www.ngowatch.org. Accessed on 10 December 2004.
6. Large multinational agribusiness companies did not feature as prominently as NGOs in the financial network. Yet they were supporting local research institutes and local companies through technology transfers and public–private partnerships.
7. http://europa.eu.int/comm/trade/issues/bilateral/data.htm.
8. http://www.monsanto.com/monsanto/layout/our_pledge/default.asp.
9. In these surveys the selected stakeholders were political actors who were, via a separate survey with key informants, identified as playing a significant role in national agricultural biotechnology debates. It was assumed that such persons would be well informed and would, therefore, have considerable influence on public opinion. This approach made it possible to conduct a survey on public attitudes in spite of a low level of public awareness of the subject.
10. See http://www.isaaa.org/kc/CBTNews/2006_Issues/Jan/Briefs_34_Highlights.pdf.

Part IV GMO Politics in the Developing World

9

The Multilevel Governance of GM Food in Mercosur

Kathryn Hochstetler

Will the Mercosur countries (Argentina, Brazil, Paraguay, Uruguay) produce, consume, and trade genetically modified (GM) food? This seemingly straightforward question in fact illuminates many of the ongoing debates about the nature of governance in the 21st century. It is not news that countries choose and implement policies under the influence of many actors both above and below the level of the nation-state. What is less well understood is how those different influences interact. In this chapter, I draw on two parallel academic literatures, on multilevel governance and global commodity chains (GCCs), to explore those interactions. Each of these approaches seeks to explain how multiple layers of decision-making relate to each other, with the former focusing on political and the latter on economic decision-making. Engaging both literatures allows a fuller view of the ways the processes interact to produce particular substantive outcomes.

The concept of multilevel governance has its origins among scholars of the European Union (EU), who developed it to understand supranational forms of governance that are taking the place of national political authority (Beyers, 2002). In this first form, it has much in common with the concept of federalism, with its formally separate levels of government. A second version of multilevel governance, used here, moves to a conception

> in which the number of jurisdictions is potentially vast rather than limited, in which jurisdictions are not aligned on just a few levels but operate at numerous territorial scales, in which jurisdictions are task-specific rather than general-purpose, and where jurisdictions are intended to be flexible rather than durable (Hooghe and Marks, 2003: 237).

In addition, authority is not limited to governments, but may be executed by other collective actors – hence, governance.

Vogler's (2003: 30–1) approach to multilevel governance fits well with this second conception and offers useful strategies for moving from its insights to concrete analysis of (changing) authority allocations. He suggests that scholars return to early understandings of international regimes to view the ways that governance institutions appear at different geographical levels. This means looking at each level for the norms, principles, rules and decision-making procedures that constitute authority in particular domains. In the analysis of decision-making on GM agricultural production that follows, I look for how these dimensions appear at levels from the local to the global. I also consider whether any given governance institution accepts or rejects the authority of others in shaping its own governance structure and what networks may link across levels.

For environmental governance in particular, Vogler insists on the importance of both brute (physical) and socially constructed facts. This will be represented here by consideration of the ways GM agricultural crops – in particular, soybeans – are embedded in GCCs. In this approach, a product is traced through multiple linked market nodes to its final consumer. The chain itself can be characterized by its length, by the kinds of interaction between different nodes, or by different patterning of the nodes (Gereffi, 1994: 96–7; Dicken et al., 2001; Fold, 2000). The approach is used here to help identify a set of crucial material components of governance that would be missed by most governance scholars, who focus on political and social facts.

This chapter puts political regulatory choice(s) at its center, as the outcome of interacting governance institutions at multiple scales and the possible instigator of a new or transformed soybean commodity chain. It is organized by the various governance sites relevant to decision-making in Mercosur about the production, consumption, and trade of GM soybeans, mostly presented in order of decreasing scale. The chapter begins with the existing global soybean commodity chain. It moves next to the inter-state level, focusing on two sites of governance: debates about trade in GMOs at the World Trade Organization (WTO) and negotiations about the Cartagena Protocol on Biosafety. The consumption end of the commodity chain appears next, with a contrast between Western Europe and China as prominent markets for soy products. At the national level, the chapter focuses on domestic decision-making on GM soy in Argentina and Brazil, responsible for about half of global soybean production in the 21st century. National decision-making in Brazil cannot be separated from the influence of subnational actors including farmers and anti-GM activists, although the subnational level is less important in Argentina. Finally, the chapter moves to the regional level of the Mercosur common market area of South America.

The global commodity chain for soybeans

In general, the most important principles and decision-making procedures of GCCs are grounded in economic market logics. Their domain is the process of production and consumption of a particular commodity. Bender and Westgren (2001: 1350) describe five major markets that link the six nodes of an agri-food commodity chain from biotechnology firms to consumers: 'One market links biotechnology [seed] firms to farmers, another links farmers to grain handlers, a third links grain handlers to processors, a fourth links processors to food manufacturers, and a fifth market links food manufacturers to consumers at retail.' These markets are potentially separate from each other in terms of the actors involved and their territoriality (Bender and Westgren, 2001: 1350–1), with the chain able to cross national and firm boundaries at any of its nodes. The primary decision-making procedures in setting up and operating GCCs are market-driven, but state regulation has also played a role in determining the links.

At present, the most contentious regulatory issues surrounding agricultural GCCs focus on possible risks from biotechnology: do GM products and processes embody risks for human or environmental health and how should such risks be handled in the absence of complete scientific evidence on the subject? Both national and international regulation of GM food have been oriented around the precautionary principle, which 'not only requires that decision-makers account for scientific uncertainty, but also provides that they can legitimately invoke scientific uncertainty to restrict a product or activity' (McAllister, 2005: 157). A survey of the biotechnology regulation of 16 countries found that only the United States presumed its safety unless risk was proven scientifically, while the other 15 presumed various levels of risk (Yu III, 2001: 587). In this context, a central concern of this chapter is how state regulation of GM products may be shaping the various links of the soy GCCs.

As a commodity, soy has certain characteristic features of cultivation and processing. Soybeans tend to be cultivated in monocultures on large farms. A Brazilian study found that soy is the most capital-intensive of all agri-business sectors and the least likely to create jobs (Roessing and Lazzarotto, 2004). Global soybean cultivation and exports are heavily concentrated in the United States, Brazil, and Argentina, which account for more than 80 percent of the global totals of each (Lapitz, Evia, and Gudynas, 2004). The soy GCCs are organized around the assumption of a homogeneous product. The first handlers typically mix soybeans of many varieties from many growers in large batches. Subsequent processors then crush them into soybean oil and meal, with low-margin profits dependent on large volumes (Bender and Westgren, 2001: 1355–6).

One particular genetic modification of soy, Monsanto's Roundup Ready (RR) soy seed, has come to dominate the GM segment of the soy GCCs. RR soy is resistant to the herbicide glyphosate, which Monsanto manufactures and sells under the brand name Roundup, so fields of RR soybeans can be easily weeded with applications of this herbicide. The environmental impact of this cultivar is mixed. On the one hand, RR soybeans require an average of 2.3 sprays versus 1.97 for conventional soy and the amount of herbicide needed more than doubles. On the other hand, glyphosate is in a lower class of toxicity than the agricultural chemical (atrazine) it replaces (Trigo and Cap, 2003: 92). In addition, the use of RR soy allows no-till techniques which help conserve soil but in the long term reduce soil quality (Chudnovsky, 2004). RR soy is among the set of GM crops whose modifications affect the production process early in the chain, but result in a product which is physically identical to the eye of the consumer – although not necessarily in the *view* of the consumer.

There is an ongoing struggle over the social construction of GM soy and whether it should be marketed as identical to other soybeans. Since the existing soybean supply chain assumes product homogeneity, any effort to sort out an identity-preserved supply chain for GM soy – like that which exists for organic soy – would require a new commodity chain. Three quite different actors would currently like to establish procedures for preserving the identity of GM soy, but for different reasons and up to different lengths of the supply chain.

Monsanto and other bioengineering companies need to preserve the identity of GM seed through the first two markets of the commodity chain if they are to make money from the products. In the first market, they are selling seed to farmers, preferably through contracts that require them to be able to do so every year. Bioengineering companies also frequently want to test for the presence of their patented genetic materials in the next market too, to catch farmers who might have saved seed or acquired it from third parties. This is a larger issue for soybeans than for some other bioengineered seeds as the characteristics of RR soy consistently reappear in succeeding generations of seed. To this end, Monsanto and other companies have developed sophisticated technologies for identifying the presence of GM crops. After these first two markets, biotechnology companies argue strenuously against any further labelling.

National governments also have interests in identifying GM products that stem from two inter-state forums. Negotiations in the context of the WTO have underlined governments' responsibilities to protect the intellectual property of biotechnology companies, and most national governments have developed legislation that enables the kinds of private claims just outlined. In addition, however, signatories of the Cartagena Protocol on Biosafety agreed to various procedures that require them to track and report the presence of GM products through the chain link that exports their products to other signatories.

Finally, many consumers want GM products tracked through every link of the soybean GCCs to its end in retail consumption – in order to avoid such consumption altogether. Governments have responded to their concerns by requiring different kinds of labelling of GM products in nearly all the developed world and in many developing countries (Baumüller, 2004). The physical exigencies of separation of GM soy (Harlander, 2002), however, ultimately require a wholly separate supply chain to meet consumer expectations. Such separation needs to be maintained in fields, farm equipment, processing plants, the transportation infrastructure, and throughout the commodity chain. Cost estimates suggest that this will add 6–50 percent to the cost of soy products depending on the product and the level of tolerance (Schaper and Parada, 2001: 29).

Two inter-state forums: trade and biodiversity

Two inter-state forums have devoted time to establishing provisions that affect trade in GM products. Trade issues are the special domain of the WTO while the negotiations over the Convention on Biological Diversity and its Cartagena Protocol on Biosafety assess the issue through the lens of biodiversity. This section considers the basic principles and provisions of each and briefly assesses their compatibility.

The WTO is well-known as an inter-state organization that is among the most closed to non-state participation. Inside it, states have engaged in contentious negotiations about how to implement the WTO's overarching principle of promoting open markets. Strong enforcement provisions have meant a second set of more specialized decision-making institutions in the various dispute resolution committees of the WTO. The 1994 Uruguay Round of the General Agreement on Tariffs and Trade that established the WTO embodied such recent trade principles as equal treatment of member states and non-discrimination against imports. The Uruguay Round also added the Agreement on Trade-Related Aspects of Intellectual Property Rights (TRIPS) which established guidelines that limited the amount of national variation on intellectual property rights and strengthened global expectations of property rights for genetic modifications (Ogolla, Lehmann and Wang, 2003).

While the WTO's overarching norm is for trade promotion, its member states have negotiated several provisions that acknowledge competing norms and establish procedures for considering them. The most important with respect to GM products are the Sanitary and Phytosanitary (SPS) Agreement that allows national regulation to protect human, animal, and plant life or health and the Technical Barriers to Trade (TBT) Agreement, which provides guidelines for packaging, marketing, and labelling (Ogolla, Lehmann and Wang, 2003; Yu III, 2001). The SPS agreement insists on the use of scientific standards for risk assessment, while the TBT agreement stresses that any guidelines under its purview must not present unnecessary barriers to trade.

A second inter-state domain opened up with negotiations that led to the 1992 Convention on Biological Diversity. The Convention's first protocol, the 2000 Cartagena Protocol on Biosafety, directly addressed the issue of transboundary movements of what it calls living modified organisms (Bail, Falkner and Marquard, 2002). The overarching principle of this set of negotiations was the protection of biodiversity, with the precautionary principle an accompanying norm for assessing the risk of imported GM products for national biodiversity. Many negotiators of the Protocol, including Argentina and Brazil, sought broad congruence with the trade-promoting principles of the WTO. This choice required a partial departure from the Convention as its norm of national sovereignty over genetic resources was contradicted by the strong private property rights negotiated as part of the later TRIPS agreement (Rosendal, 2001).

The final language of the Cartagena Protocol left its formal relationship to the trade principles unclear (Bernasconi-Osterwalder, 2001: 713–4). Nonetheless, legal scholars agree that the actual content of the Cartagena Protocol is broadly compatible with existing trade agreements, with one assessment placing all of its provisions in the category of 'General compatibility, but possible ambiguity and/or uncertainty' (Ogolla, Lehmann and Wang, 2003: 123. The clearest potential incompatibility is in the handling of risk assessment and the precautionary principle, with the SPS agreement requiring a short and active period of risk assessment while the Cartagena Protocol allows for longer indeterminacy (Baumüller, 2004; Bernasconi-Osterwalder, 2001; Yu III, 2001; see also Chapter 11).

In practice, much of the recent contention about the fit of the trade and biosafety principles has focused on issues more closely connected to the TBT agreement. The contentious issue of how to identify GM content in commodity shipments was put off until the first Meeting of Parties in February 2004 (Falkner and Gupta, 2004). By this point, most Protocol signatories had adopted varying provisions requiring labelling of some kind, which spurred an effort among exporting countries to try to reach a uniform and less stringent set of requirements. The Parties have been unable to reach agreement at their first two meetings in 2004 and 2005. Argentina and Uruguay have sided with the pro-GM position of the US and Canada on these issues, while the Brazilian role has been unpredictable, for reasons discussed below.

In summary, both the biosafety and trade inter-state agreements limit the range of nation-state policies, while still allowing for a variety of national choices on how to assess the risk of biotechnology and how to protect environments and human health. The two inter-state domains are potentially compatible, but particular variants of biosafety regulations might not fit with WTO prescriptions. Neither is a hard-constraining level for nation-states yet, although ongoing negotiations may limit either GM-promoting or -restricting choices in the future.

Consuming interests

Consumers are at the very end of global production chains and have exercised considerable influence over the shape of those chains and the inter-state negotiations about them. The Mercosur countries share the same two largest export markets in the EU and China and so they are the focus here. There have been some fairly dramatic recent shifts in who imports soy from whom. From 1993 to 2002, the EU turned away from the GM soybean producers of the United States (7695 to 5910 million tons) and Argentina (1929 to 1176 million tons or below) and turned their markets over to largely GM-free Brazil (3192 to 9196 million tons). Over the same years, all three exporters gained a major new market in China, going from essentially no exports in 1993 to over 4000 million tons of beans each for the United States and Brazil in 2002, with Argentina somewhat lower (Pereira, 2004: 30). This section traces the governance structures and principles relevant to soy imports in the EU and China.

The EU has developed the most stringent regulation of GM products in the world and has become one of the strongest proponents of collective regulation of biosafety issues in global negotiations (Bail, Decaestecker and Jørgensen, 2002). These positions can be traced to the influence of strongly mobilized environmental and consumer groups and a bedrock of sceptical public opinion (Bernauer, 2003: 78; Newell, 2003c: 65; Purdue, 2000). In this node, the most effective governance structure is in the public sphere. A succinct version of the principle orienting this governance site comes from the 95 percent of European respondents who told Eurobarometer in 2001 that 'they must have the right to know whether or not GE products are in their food, and they must have the right to choose' (Bernauer, 2003: 74). European state policy-makers have adopted policies that follow this norm and defended it in international forums, but they are not the central decision-makers. In fact, the formal policies are redundant due to the number of European retailers who had already agreed not to sell GM food products or at least to label them (Bernauer, 2003: 88). The United States has challenged the EU's policies at the WTO, but such strategies are likely to backfire as they overlook the central role of trust issues in markets (Bender and Westgren, 2001).

The Chinese market presents an interesting contrast in governance type. Since 2001, its regulations governing trade and labelling of GM food have been close to the European model (Chapter 10 in this volume; Newell, 2003b: 5; Yang, 2003). There is a major difference between China and the EU, however, in that China's policies are those of the Chinese state and reflect its concerns. The Chinese population's preferences on GM food are largely unknown and likely to reflect state information in any case. Thus for the Chinese import market, the crucial governance structure is the formal decision-making structure of the state and its preferences (Newell, 2003b: 23). The Chinese government joins a major commitment to biotechnology development and production *with* consideration of social concerns in its biotechnology choices *and* the development of extensive regulation of that development. Newell notes that China has overtly accepted European norms on GM and used them to justify its stringent internal regulations, but then crafts the regulations in such a way that domestic producers are benefited over international ones (Newell, 2003b: 38).

In summary, the consumption node takes on quite different governance forms in the EU and China, despite superficially similar government policies. Consumers and NGOs shape the larger EU governance structure to be strongly against GM products, whatever the outcomes of formal government agreements. In contrast, the Chinese state dominates its node, where it takes varying positions on biotechnology but also buys a rapidly increasing share of global soybeans of all kinds.

Argentina

Argentina offers comparatively straightforward answers to the questions posed in the opening paragraph of this paper. It will produce and consume GM agricultural crops, and has since 1996. Argentina trails only the US in every measure of commitment to growing such crops, and has been a steady partner in the international negotiations promoting their free trade. The decision-making procedures relevant to GM production with Argentina have remained unusually concentrated in the regulatory capacity of the national executive. Nonetheless, that monopoly has come under serious challenge, first by organized civil society and now by Monsanto. Both of these challengers have put pressure on the Argentine government by drawing on other levels of governance. The more serious challenge by Monsanto reflects Argentina's incomplete adoption of emerging global property rights principles.

GM soy is Argentina's most extensive GM crop, comprising almost 90 percent of the 12 million hectares planted to soy in 2001–02 (Trigo and Cap, 2003: 87). The production of soy has increased dramatically, rising from 10.9 tons and about one quarter of total agricultural production in 1990–91 to 35.0 tons in 2002–03, a figure that was nearly half of all of Argentina's agricultural production (Chudnovsky, 2004: 16). Argentina exports nearly all (98 percent in 2003) of its soy as beans or other products; collectively, they have come to represent about 20 percent of Argentina's total exports (Lapegna and Domínguez, 2004: 12). Argentina's experience with GM soy is in many ways archetypal of a particular kind of export agriculture model, but it also has unusual characteristics that mean its comparative success there will not be readily duplicated for other GM crops or other cultivators of GM soy.

The roots of the explosion of GM soy production in Argentina date to 1991, when the then president Menem transformed the Argentine economy, opening it up to global and national market forces. Menem endorsed a vision of export-driven growth that depended heavily on the development of a globally competitive modern agricultural sector (Bisang, 2003; Chudnovsky, 2004). His government established a regulatory framework for biotechnology that same year, creating the National Commission of Agricultural Biotechnology (CONABIA) inside the Ministry of Agriculture, which promptly authorized the first field trials of GM soy. CONABIA approved the production and sale of GM soy on a commercial scale in 1996. Argentina's early adoption of GMOs meant that it escaped the scrutiny other governments would face – at least in the early stages. While there was societal opposition to Menem's market-oriented policies, it did not center on the agricultural policies and even less on the beginnings of GM production. The global anti-GM movement emerged only in the mid-1990s (Purdue, 2000) and came later to Argentina. By the time Greenpeace, the Network for Rural Reflection, and other organizations formed an Alert Network on Transgenics in 1999, 75 percent of the Argentine soy crop was already GM.

The most unusual aspect of Argentina's extensive cultivation of what are essentially Monsanto's RR soybeans is that Monsanto does not have an exclusive patent for its seeds in Argentina. In an early agreement, Monsanto licensed an Argentine firm, Asgrow Argentina, to have access to its RR gene. The multinational firm Nidera acquired Asgrow and the gene access, and disseminated the seed in Argentina. When Monsanto asked to patent its RR soybean seed there, it was denied an exclusive patent on the grounds that the gene had already been released. In addition, Argentina adheres to the 1978 version of the UPOV treaty, which allows farmers to save seed for their own purposes. Since the herbicide-resistant gene in RR soy comes true from seed, farmers are able to reseed without paying royalties to Monsanto – and a thriving black market sells to new cultivators (Chudnovsky, 2004: 19). For these reasons, the price of RR soybean seed in Argentina is well below global market rates. Farmers benefit further from the fact that Monsanto's patent on the Roundup version of glyphosate ran out just before GM soy was approved. The end result is that farmers accrued some 87 percent of the added economic benefits from adopting GM soy in Argentina from 1996–2001, while suppliers accrued only 13 percent (Chudnovsky, 2004: 21). The percentage benefit shares are virtually reversed for GM cotton and corn crops in Argentina, and farmers have been much slower to adopt them (ibid). The success of GM soy for farmers in Argentina is clearly unlikely to transfer elsewhere – at least not legally.

The debate on GM food is only slowly beginning to open up to additional points of view in Argentina. The Secretariat of the Environment has finally secured a place in CONABIA, under Menem's successor, and had two participants in 2005. There are still no representatives of consumer or environmental groups on CONABIA, although the membership list includes six representatives of seed companies, including Monsanto, Pioneer, Dow, and Bayer. Organized civil society has had to make its own way into these debates. As noted, there has been an anti-GM network since 1999, but it has encountered mixed results. The Network for Rural Reflection proposed asking the Public Defender for a moratorium on all transgenic production in 1999,[1] but no action was taken. Two years later, Greenpeace mounted a campaign against a legislative project that would have codified the existing control of the Ministry of Agriculture, and claimed credit for its failure in

November 2001. One part of Greenpeace's argument, supported by the Argentine Ministry of Foreign Affairs, was that this law would be incompatible with the Cartagena Protocol of Biosafety, signed but not ratified by Argentina.[2] Greenpeace's Argentina campaign is part of a global anti-GM campaign by the international organization.

The Argentine population has not openly taken sides on GM issues. A survey done in 2001 for the Argentine Association of Seed Producers found little understanding of genetic modification. For example, almost half of the interviewees (46.1 percent) believed that if a person eats GM food, that person's own genes are altered. At the same time, they believed (78.4 percent) that hunger and poverty are the primary global problems and (67.7 percent) that science could solve those problems (Días, 2001).

Argentina's severe economic and political crisis in 2001–02 raised the paradox of the world's 8th largest food producer exporting soybeans for animal feed while half of its population dropped below the poverty line and significant numbers went hungry (Lapegna and Domínguez, 2004: 14). Large sectors of the population and the present president Kirchner have raised many questions about the value of the economic model Menem brought to Argentina. At the same time, export agriculture has been one of the few bright spots in a shaken economy and continues to generate significant national revenues. In this context, it is perhaps not surprising that Kirchner's administration has continued to support GM production. For many of the same reasons, however, the administration is also strongly resisting Monsanto's recent efforts to regularize its royalties for RR soy.

As Argentine politics returned to near normal with Kirchner's election in 2003, Monsanto began to press to receive credit for its genetic innovation in the second market of the soybean commodity chain, where farmers sell their crops to handlers. Monsanto stopped selling its own seed in Argentina in December 2003, pushing the government into negotiations about additional compensation. Such a bill promised at the end of 2004 did not materialize, following strong farmer resistance (*World Intellectual Property Report*, 2005). Monsanto shifted venues in early June 2005, filing lawsuits in Denmark and the Netherlands – where it has a patent – to collect royalties on imported RR soy from Argentina. These trials are beginning in September 2005 and could result in European courts effectively setting compensation levels for Monsanto in Argentina. The Argentine government has strongly opposed such an outcome, convoking its regional neighbors to stand with it (see Mercosur section, below). The Secretariat of Agriculture is asserting the jurisdiction of its national regulatory framework and citing Argentina's national interest in the issue. At the same time, it is drawing on global principles, criticizing Monsanto's position as anti-competitive, and pointing out that Monsanto itself has insisted in international forums that RR soy and the conventional soy product are substantially equivalent. In meetings with the US Secretary of Agriculture, Argentine Secretary Campos has also reminded the US of Argentina's role as its strategic ally in global negotiations.[3] The outcome of this conflict will be potentially precedent-setting and will depend in part on Argentina's neighbors, especially giant Brazil.

Brazil

The decision-making procedure on GMOs in Brazil looked briefly like Argentina's before exploding into a decade-long political debate that involved all of the three branches of national government, lower levels of governments, and non-state actors. Brazil has very recently (March 2005) passed national legislation that facilitates legalization of commercial GM crops. Even so, the principles, norms, and rules governing GMOs there are still in flux, affecting global as well as national GM food politics.

By the mid-1990s, a large-scale, export-oriented agriculture model was well-established in Brazil, as the Cardoso administration began to put the next pieces in place to follow the world into GM agriculture. The 1995 Biosafety Law established the basic regulatory framework and created a National Technical Commission on Biosafety (CTNBio), with participants from a number of government ministries and biotechnology scientists as well as consumer, worker safety, and biotechnology industry representatives. CTNBio was given sole authority to evaluate the safety or risk of transgenics and to regulate their use. A new patent law in 1996 then generally established global-standard intellectual property rights. As in Argentina, however, Brazilian law follows the 1978 UPOV agreement, allowing farmers to save seed (Palaez and Schmidt, 2004: 243–4). The Ministry of Agriculture has consistently supported this regulatory framework, as do many growers, as it represents a broadly permissive or even promotional stance on biotechnology (Paarlberg, 2001: 69–74). Like Argentina's CONABIA, CTNBio moved promptly to authorize field trials in 1996 and then commercialization of GM crops, beginning with Monsanto's RR soybeans in 1998.

At this point, the stories of these neighbors began to diverge. Brazil has more active civil society organizations on environmental and consumer issues than does Argentina, and they have stronger legal and institutional tools to use to influence government policies (Hochstetler, 2002). From the outset, the consumer representative on CTNBio, the Institute for Consumer Defense (IDEC), had an inside view of the approval process. When the Commission was poised to approve RR soybeans, IDEC resigned its seat and moved to the courts, arguing for a restraining order. IDEC was eventually joined in its legal and political challenges by Greenpeace and an array of other social movements as well as by the Ministry of the Environment and its implementation arm, IBAMA. They drew on an existing social movement network from the 1970s that had opposed the Green Revolution's agricultural model (Palaez and Schmidt, 2004). The coalition's main argument was the constitutional one that the creation of CTNBio violated constitutional distributions of authority to the Ministries of Health and Environment. In particular, the Ministry of the Environment has the right and responsibility to determine when environmental impact assessments are required, which the 1995 law gave to CTNBio for biotechnology cases (IDEC, 2005).

From 1998, when IDEC filed its first lawsuit, until now, the politics of GM food in Brazil has been polarized around these two procedural options – CTNBio versus the Ministry of the Environment – and the presumably different outcomes for GM approval that they represent. Proponents of both sides have pursued their struggles in both the judicial and legislative branches, with wins and losses for each.

Anti-GM activists won a series of judgments and appeals to them through 2000, both on a preliminary set of actions seeking an injunction and then on the merits of the case itself (McAllister, 2005). The legal conclusion through this stage invoked the international legal precautionary principle and called for more extensive risk analysis through an environmental impact study that IBAMA would execute. In 2002, the norms-setting body of the Ministry of the Environment, CONAMA, quietly established procedures for assessing and licensing GMOs (Res. 305, 12/6/02). More recently, the court decisions have begun to turn in the GM-permissive direction. The last two decisions, at the federal appellate court level, have determined that there has been sufficient study to justify lifting the ban on GM soy. The appellate court's decisions upheld the precautionary principle, but argued that CTNBio could be the primary evaluator of risk and could do so through scientific methods other than an environmental impact assessment if it chose. Further appeal is allowed because the decision at the federal appellate level was split (McAllister, 2005).

Anti-GM activists expected that the 2002 election of the first president from the Workers' Party (PT) would aid their cause and that President Lula would replace the 1995 regulation, making the court cases moot. A PT governor in the state of Rio Grande do Sul had declared his state a GM-free zone in 1999 (Jepson, 2002; Palaez and Schmidt, 2004) and the PT itself had challenged the constitutionality of the 1995 law in court (Ação Direta de Inconstitucionalidade No. 2007). As a candidate, Lula often spoke against GMOs and the large-scale agricultural model to which they are linked, but his administration's positions have been mixed. Minister of the Environment Marina Silva strongly rejects GMOs and has argued for her agency to retain approval power over them while Minister of Agriculture Roberto Rodrigues equally strongly welcomes GMOs and supports the powers of CTNBio.

The Lula administration did replace the 1995 legislation, but not in the way expected. First, annual executive decree laws began allowing farmers in Rio Grande do Sul state and then elsewhere to sell their illegally planted GM soy, beginning with the 2002–03 harvest season. These decisions infuriated some of the PT's traditional supporters, including the very active and vocal

Landless Movement (MST). The MST opposes GM crop production, largely for its socio-economic and some environmental impacts. With a broad coalition of anti-GM social forces, it has been quite combative, even burning trial fields and occupying a Monsanto plant (Guivant, 2002). Environmental organizations broke with the government quite publicly over the GM issue, with 500 groups sending Lula a letter of personal repudiation in October 2003 (Hochstetler, 2004). These actions were designed to support Silva and her position in the administration, and the government responded by hastily revising the draft of a new Law on Biosafety which went to the National Congress at the end of the month. The proposed legislation followed civil society organizations and the then current legal decisions, and gave the Ministry of the Environment the major role in performing environmental impact assessments of GMOs before they could be widely grown and sold. Notwithstanding its public capitulation, the Lula administration then worked to undermine its own bill in the Congress.[4] Amendments in the Senate once again placed the CTNBio in charge of evaluating genetic modifications, allowed the planting of GM soy, with labelling, and added provisions for stem cell research. Physically handicapped protesters in favor of stem cell research strongly lobbied for the bill and helped it pass in the Congress (*Jornal da Câmara*, March 4, 2005: 4–5). Lula signed the bill in March 2005. Nonetheless, the Brazilian case can be summed up as another indeterminate level – with two potentially intractable and opposed sublevels. Each of them has behaved much more consistently than the national government, restricting its scope for autonomous action.

On the one hand, opponents of GMOs have shown that they have considerable political power and mechanisms to exercise it. Through the courts and through their political pressure on decision-makers, they were able to prevent the legal cultivation of GM crops in Brazil for almost 10 years, and their opposition is continuing. In June, the Federal Prosecutor General filed a suit of unconstitutionality (ADIN No. 3526) with the Federal Supreme Court, following requests made by the Green Party and IDEC in May. Greenpeace, IDEC, and the National Association of Small Agricultural Producers have filed *amicus curiae* briefs in the case, as has the National Biosafety Association. The case is still in process.[5] IDEC has also been promoting a citizen campaign for labelling and has legally challenged the Brazilian Association of Food Industries' opposition to labelling. Whatever the legal status of GMOs, this is a coalition ready to monitor agricultural production and to pounce on any evidence of negative social consequences. It has already moved to document the growing impact of soy production on the environment in the Amazon region in particular (for example, WWF, 2003).

The anti-GM coalition is allied both directly and indirectly with foreign consumers, environmentalists, and anti-GM activists. Thus, when the Brazilian food exporter Grupo Maggi was singled out by *The Economist* magazine as facing pressures from European consumers worried about the corporation's contributions to deforestation in Brazil, WWF recruited Grupo Maggi for a roundtable discussion on sustainable soy. (Bruno Maggi is governor of Mato Grosso state, where half the deforestation of 2003–04 took place.) GM opponents also have some broader political support in Brazil. A survey commissioned by Greenpeace in the early 2000s found quite high levels of GM rejection. In the seven Brazilian state capitals surveyed, 73.9 percent of respondents believed that consuming GM food was risky for their health and 81.9 percent believed that GMOs should not be planted in Brazil. Greenpeace-Brazil has 61 businesses on a 'Green List' who have guaranteed GM-free products, with seven of them joining after the 2005 bill was passed.[6]

On the other hand, one of the few constants in this story was the willingness of Brazilian farmers to plant GM soybean seed year after year, even illegally. Already in 1999, 400,000–750,000 of the 2 million hectares of soybeans planted in Rio Grande do Sul were estimated to be transgenic (Paarlberg, 2001: 31). A 2003 estimate placed the figure at 80 percent for the same state (CLAES, 2003). With Rio Grande do Sul the center of such illegal plantings, it is worth noting that it is also the state with by far the largest number of small soy farms and with more than half of Brazil's entire population of people who work with soy (Roessing and Lazzarotto, 2004: 32, 35). This would have made a legal crackdown difficult logistically and politically. Ironically, the legalization of GM soy may prove to limit farmers' use of the product. With Monsanto poised both to charge for seeds and to levy a fine at processing plants on old seed, sales are down across the country. The company has been forced to reduce its proposed price for seeds from R$0.88/kilo to R$0.50 (*Plataforma Soja*, August and September 2005). Monsanto's negotiations with Brazilian seed multipliers, seed organizations, and growers have all been contentious, even though the Brazilians freely acknowledge its right to collect some fees for its technology. The Lula administration, mired in corruption charges, has left these negotiations largely to private actors.

Other private actors have also weighed in to push Brazilians against growing GM soy. The British Retail Consortium has issued such a call, saying it needs non-GM soy to meet consumer demands.[7] The conventional soy GCCs are facilitated by decisions like those of the governor of Paraná state to keep his state's port devoted to conventional soy, still 94 percent of its grain; a private side facility will handle GM soy (*Plataforma Soja*, September 2005).

For more than a decade, domestic Brazilian positions on GMOs have been deeply divided, shaping its schizophrenic international participation. Brazil's delegations to environmental conferences themselves are always multi-vocal, with representatives of environmental agencies grouped with agriculture, science and technology, and trade agencies (Lisboa, 2002; Nogueira, 2002). High-level interventions have had to settle intra-delegation disagreements, with a phone call from the then Minister of Environment

Sarney to the then president Cardoso achieving Brazil's adherence to the Cartagena Protocol in 2000, while a similar call from Lula's Chief of Staff José Dirceu sunk the labelling negotiations in Montreal in 2005.[8] Brazil's future international participation is likely to continue to be unpredictable, given the competing undertows of domestic GMO politics and economics.

Mercosur

Mercosur, the Common Market of the South, has operated as a partial customs union since 1995, creating very few supranational institutions and working primarily as a series of regional gatherings of existing institutions (Hochstetler, 2003; Phillips, 2001). The national presidents form the highest decision-making body for Mercosur, with representatives of the national economic and foreign affairs ministries overseeing most of Mercosur's activities. It is dominated economically and politically by Brazil and Argentina, with Paraguay and Uruguay as smaller partners. Regular political and economic crises in the region have made Mercosur's very existence a continual issue (Gomez Mera, 2005). With the Brazilian and Argentine cases already introduced individually, it will not be surprising that the GMOs issue has been divisive for the region.

The Uruguayan and Paraguayan policies on biotechnology deserve brief notation first. Uruguay broadly matches the Argentine pattern: 90 percent of its soy acreage is planted to RR soy (although its total area is less than some individual farms in Brazil and Argentina) (Schaper and Parada, 2001: 45). Uruguay has also negotiated in global biosafety forums with Argentina as part of the US-led group of agricultural exporters. In contrast to Argentina, however, Uruguay's decision-making structures for GMOs give authority for licensing and risk assessment to a collection of ministries that include the economic, agriculture, health, and environmental ministries – and Monsanto holds the patent for RR soy there. GM products in Paraguay have been 'neither prohibited nor approved' (CLAES, 2004). For several years, the Ministry of Agriculture issued annual prohibitions on commercialization of GMOs, citing the Convention on Biological Diversity's precautionary principle (Schaper and Parada, 2001: 42). Even so, an estimated 80 percent of Paraguay's soy crop was already transgenic in 2004 (CLAES, 2004), with seeds coming illegally from Argentina.

Biosafety issues were most extensively discussed in Mercosur well before most of the events in this chapter. In 1996, Mercosur's Working Group No. 6 on the Environment (SGT6) began to work on an Environmental Protocol for Mercosur that would harmonize the very uneven national environmental legislations (Hochstetler, 2003: 7–12). With regional environmental leader Brazil producing the initial draft of the document, the first version had a substantial section on biosafety (SEMA-SP, 1997: Anexo 1). A year of discussions within SGT6 and consultations with other national government agencies and civil society cut the draft's references to the Convention on Biological Diversity and its call for extensive public discussion on GMOs. The new version kept an overall orientation that perceived biotechnology as risky for the environment and required full assessment of the potential risks of GMOs throughout the entire process of production (SGT6/Acta 06/97, Anexo 10).[9]

SGT6 raised this draft to the Common Market Group (GMC – of national economic and foreign ministries) for possible adoption in June 1997 and then waited another year to receive a firm rejection. Most of the objections within the GMC came from Argentina, and centered on the biosafety clause (Hochstetler, 2003: 20). In the end, the Environmental Protocol was heavily revised under the foreign ministry guidance and even downgraded to the status of a Framework Agreement on the Environment. As adopted in 2001, the Framework Agreement has no section on biosafety issues at all, although biosafety is included in a long list of issues in an Appendix of areas for possible future discussion (CMC/Acta Nº 02/01).

In October 2004, the Agricultural Subgroup (SGT8) reported to the GMC that it saw little possibility for coordination of regional agricultural policies in the near future and that it would focus on reducing whatever barriers to trade might exist instead (GMC Ata Nº 03/04). The change in Brazilian legislation in 2005 seemed to open up the possibility for cooperation. Just a month later, the Argentine government called a special meeting of the Ministries of Agriculture of the Southern Farm Council to drum up regional support for its position against paying Monsanto royalties on soy crops rather than seeds. At the meetings, the ministers of Brazil and Paraguay apparently agreed with the Argentine position, a rare point of unity within the trade bloc. Just a few days later, however, the Brazilians and Paraguayans back-pedalled, both citing their distinct national legislation and noting that their private sectors were reaching their own agreements with Monsanto. Uruguay had not even sent a first-line negotiator.[10] The national level thus continues to dominate the regional one on GMOs in Mercosur.

Conclusion

This chapter shows quite clearly the multifaceted and still unpredictable nature of global governance of GM food and its specific forms in Mercosur. Brazil itself represents a microcosm of much of the global political debate around GMOs. It has both the irresistible force of agricultural producers determined to take advantage of a new and – to them – benign technology *and* the immoveable object of globally networked consumer and environmental activists equally determined to stop GM crop production they consider treacherous. Private biotechnology companies are present and pressing their claims hard to a state that is only partially able to respond with authority, and is itself conflicted. Even Argentina's clearer position shows the incompleteness of its adoption of global trade principles.

In theoretical terms, the article clearly shows the usefulness of a focus on multilevel governance, which shows the many kinds and principles of governance at work on this issue. The physical components of the soy commodity chain, the reluctance of European consumers to eat GM food, and the determination of producers to grow them are comparatively fixed nodes in this process. Beyond those nodes, and especially in the way they fit together, political constructive efforts are also a key force whose final impact is still being worked out.

Notes

1. www.laneta.apc.org/ogt/redalerta.html.
2. www.greenpeace.org.ar.
3. SAGPyA (Secretariat of Agriculture) Press Releases 31 March 2005, 1 July 2005 and 14 September 2005, online at http://www.sagpya.mecon.gov.ar/. See also Pardo (2005).
4. Interview with Marijane Lisboa, São Paulo, 12 August 2005. Lisboa was the Secretary of Environmental Quality in the Ministry of the Environment in 2003–04.
5. Cases can be tracked on the Court's website: www.stf.gov.br/processos/.
6. See www.greenpeace.org.br and www.idec.org.br.
7. See www.foe.org.uk.
8. Interview with Lisboa.
9. The various Mercosur Actas cited parenthetically in this section can be found at www.mercosur.org.uy.
10. See www.abrasem.com.br/informativo/anexo/Inf41.doc. Abrasem is the Brazilian national seed organization.

10
Internationalising Biotechnology Policy in China

Robert Falkner

Introduction[1]

China's agricultural biotechnology policy has undergone a profound transformation over the last decade. From the 1980s onwards, China was an enthusiastic promoter of modern biotechnology and became the world's first country to grow a genetically modified (GM) crop on a commercial scale. More recently however, the country has tried to shore up its environmental risk regulations for genetically modified organisms (GMOs) and halted the authorisation of new GM crops in the late 1990s. The Chinese leadership has proclaimed that all benefits *and* risks of genetic engineering need to be considered before more GM products can be approved. This move towards greater precaution in agricultural biotechnology supports the view that environmental concerns have moved up the political agenda in China. At a minimum, it suggests at least a partial break with the unbridled technological optimism that characterised communist rule in the past.

Unlike in Europe, where consumer hostility towards GM food has forced a strengthening of the European Union's biotechnology regulations, the shift in China's policy is not the direct result of societal pressure. Indeed, much of the recent focus on sustainability issues in Chinese politics, however shallow it may be, can only partly be explained with reference to the rise of an environmental movement. The nature of the autocratic political system and the weakness of civil society combine to constrain environmental activism and its political effects. Instead, we need to look for other explanatory factors. Ideological debates and bureaucratic politics within the core state play an important role here as has been noted in other policy areas (Fewsmith, 2001b). But as this chapter argues, change in China's environmental policy is also influenced by international forces, and particularly so in the case of GMO regulation.

China's integration into the international system has produced two dynamics that impact on domestic environmental policy. The first dynamic is the result of China's ongoing international socialisation (Kent, 2002).

Participation in international environmental regimes provides mechanisms for transmitting global environmental concerns and norms into the domestic political arena. The second dynamic originates in the myriad of global links that tie China's domestic economy to international markets (Moore, 2002). Countries with higher regulatory standards can export these by imposing environmental restrictions on imports from China, thus producing an economically motivated 'trading up' effect. Where both dynamics work together and reinforce each other, China's global integration can become a powerful source for environmental policy change.

The growing internationalisation of domestic politics has spawned a large research literature in international relations and globalization studies. Two principal arguments have dominated the debate. The first, based on rationalist assumptions and widely employed in international political economy, views internationalisation as a process through which the incentives for governments, firms and socio-economic groups change within societies (Keohane and Milner, 1996; Simmons and Elkins, 2004). This change in incentives generates new policy preferences and new coalitions of actors, potentially leading to domestic policy change. The second, based on constructivist assumptions, sees states as embedded in a wider framework of international norms and institutions, which affect the formation of preferences, and even identity, of domestic actors (Finnemore and Sikkink, 1998; Risse, Ropp and Sikkink, 1999). Through mechanisms of socialisation and social learning and the operations of transnational networks, international norms work their way into domestic politics and contribute to policy change. Much of the debate in international relations has centred on the question of which of these two perspectives better explains international–domestic linkages. More recent work, however, has begun to build bridges between rationalist and constructivist logics and has concentrated on identifying and elaborating the causal mechanisms that link international forces and domestic policy change (Zürn and Checkel, 2005). This chapter seeks to contribute to this effort, providing an empirical case study that illustrates the dual nature of the internationalisation of environmental policy change in China.

It is argued below that both dynamics – change in domestic preferences through economic globalization and the adoption of environmental norms through transnational networks and concern transfer – have helped to promote an environmental safety agenda as part of China's agricultural biotechnology policy. The country's growing international integration has exposed environmental policy-making, in often subtle and indirect ways, to a range of international forces and influences. Despite a comparatively high degree of state control over the domestic political agenda and a long history of resistance against globalization (Hughes, 1997), China has been unable to isolate itself from the international GM food controversy.

The chapter proceeds in four steps. The first section provides an overview of the evolution of China's biotechnology policy and the recent shift towards a more precautionary stance on biosafety. The second section analyses the effect that China's international integration, and particularly participation in the international biosafety regime, has had on domestic policy. The third section examines the impact of economic globalization in agriculture and the 'trading up' effect produced by international trade restrictions on GM crops. The final section summarises the findings and looks at the challenges facing China's biotechnology and environmental policy.

Shifts in China's biotechnology policy

Genetic engineering of plants represents a revolutionary technological change in agriculture. Unlike traditional forms of plant and animal breeding, recombinant DNA techniques enable researchers to directly manipulate the genetic composition of target organisms. By inserting, removing or altering genes, genetic engineering produces much faster, and more targeted, forms of genetic change, which can also occur across species boundaries. The nature and relative novelty of this technology, which has been developed since the 1970s and commercialised since the 1990s, led to an intensive debate over the desirability and safety of GMOs. Advocates of the technology argue that GM plants can improve food quality and increase agricultural productivity; they see GM crops as key to ensuring food security in developing countries. Critics point to risks for human health and the environment, including threats to biological diversity; they question whether agri-biotechnology, which is capital-intensive and dominated by Northern multinationals, is a suitable and socially acceptable technology for the developing world (for an overview of this debate, see Falkner, 2004a: 249–52).

China's biotechnology strategy

Although agricultural biotechnology is dominated by a few multinational corporations, a number of developing countries have sought to develop their own research capacity and to adapt the new technology to domestic needs. China has taken a leading role among them. Genetic engineering has been an integral element of China's national agricultural strategy since the early 1980s. In a country preoccupied with food security and intent on catching up with Western technological advances, biotechnology's promise of increased yields, more reliable harvests and reduced chemical inputs was taken more seriously than almost anywhere else in the developing world. During the 1980s and 1990s, the Chinese state provided ever-increasing public funds for research and development, and even today the Chinese state plays a decisive role in funding and directing the country's biotechnological R&D and commercialisation.

The origins of China's biotech programme go back to the early phase of the economic modernisation programme initiated in the late 1970s. Genetic engineering, alongside computing and space technology, became one of the key areas of Deng Xiaoping's reform policy. It held the promise of not only improving farm yields but also helping China to take a leap in the global technological race (Suttmeier, 1980). The country was able to build the largest scientific basis for advanced biotech research outside the industrialised world, with over 150 national and local research laboratories in operation today (Huang and Wang, 2002). An estimated 2690 scientists were working in the field of plant biotechnology in 2003, up from 740 in 1986 (Huang et al., 2004: 7), and Chinese research institutes reported in 2002 that scientists had produced 141 types of GM crops, 65 of which were already in field trials (Huang and Wang, 2003). Despite these impressive achievements, doubts persist as to whether China has really closed the technological gap with the West. On the whole, Chinese research laboratories are better at adapting international GMO developments to local conditions than engaging in the kind of basic research that has allowed Northern biotech firms to dominate the field. Partly in recognition of these limitations, China has now somewhat reluctantly accepted the need for co-operation with biotech multinationals, although foreign direct investment (FDI) in agri-biotechnology remains a politically charged issue.

China's headlong rush into modern biotechnology proceeded largely unencumbered by any significant regulatory regime well into the 1990s. Whereas most leading biotechnology countries had created a comprehensive system of biotechnology governance by the late 1980s, China did not even begin to establish biosafety regulations until the early 1990s, long after other developing countries such as India had started to address the environmental and health concerns surrounding genetic engineering (Gupta, 2000; Paarlberg, 2001). The absence of any safety regulation for genetic engineering played into the hands of Chinese researchers who in the late 1980s were the first to grow a GM crop in commercial quantities worldwide, a virus-resistant tobacco plant (Paarlberg, 2001: 128). In 1997, 12 other GM crops were approved for field trials, of which three passed the safety tests for commercial planting (cotton, tomato and petunia). Of the GM crops approved for introduction to the market, only cotton has been grown on a large scale since 1997, with insect-resistant Bt cotton accounting for 58 percent of the total production in 2003. An estimated 5 million farmers are now using Bt cotton, including also varieties developed by Monsanto, the first and so far only multinational to sell GM seeds through a joint venture with a Chinese firm (Huang and Wang, 2003).

Shift towards a more precautionary stance on biosafety

In the late 1990s, China was set to authorise a range of new biotechnology products when suddenly in 1999 a *de facto* moratorium on new GMO releases was introduced.[2] The timing of this unofficial ban on GM crop developments – after the European Union introduced its own GMO moratorium in late 1998 and shortly before the adoption of the Cartagena Protocol in

Biosafety in January 2000 – is significant. It reflected the growing unease among scientists and policy-makers in China about the environmental risks and trade implications of domestic GM crop authorisations. Against the background of a shift in the domestic scientific debate and a growth in global anti-GM protests and GMO trade restrictions, China began to reassess its approach to agri-biotechnology.

The 1999 moratorium was widely perceived as a significant policy shift, for in the preceding years China's efforts at ensuring biosafety had been comparatively lax. The first Chinese biosafety rules that were introduced in 1993 provided a mere framework for regulation. The Ministry of Science of Technology (MOST) was the lead agency dealing with biotechnology matters at that time and established the Safety Administration Regulation on Genetic Engineering, a set of general safety rules drafted by scientists for scientists. It took 3 years for the Ministry of Agriculture (MOA) to follow this up with its Implementation Guidelines of 1996, which became the basis for authorising the commercialisation of GM crops. Just like MOST's guidelines, the 1996 regulations were informed by a desire to promote biotechnology and concentrated on scientifically demonstrated risks (Paarlberg, 2001: 129) – a position that tended to downgrade the importance of long-term and uncertain threats to human health and environment. It was only from 1996 onwards, nearly 8 years after the first GM crops had been planted in China, that Chinese authorities required a case-by-case risk assessment of new developments in agri-biotechnology. MOA has since taken the lead in regulating and authorising GM crops for agricultural production, covering all aspects of biosafety assessment. Given its close links with the agricultural sector and biotech research institutes, MOA was widely seen to favour the rapid commercialisation of GM crops (Keeley, 2003c: 16).

A shift towards greater environmental awareness was already under way when in 2000 the allocation of regulatory authority within the state bureaucracy changed yet again. With the adoption of a new national seed law, the final managerial authority over all new GM crop varieties passed to the State Council, a central decision-making body at cabinet level. This reorganisation produced a more centralised system of GM regulation, thus acknowledging the greater political significance of biotech-related decisions. Some observers also interpret it as an attempt to take away regulatory authority from regional governments that had been able in the past to approve field trials by foreign biotechnology firms, as happened in Hebei in 1994 with the controversial decision to introduce Monsanto's GM cotton variety (Paarlberg, 2001: 132).

Reflecting greater awareness of environmental concerns and the newly created Cartagena Protocol on Biosafety, the State Council produced new and more stringent biosafety rules. The State Council's 2001 Regulation on Safety Administration of Agricultural GMOs was followed in 2002 by three implementing regulations issued by MOA, covering the areas of biosafety evaluation, import safety administration and GM food labelling. These new acts provided a more comprehensive system of risk management, for the first time covering imported GMOs and providing consumers with some degree of choice. They also signified a shift away from the previous product-based risk assessment of GMOs, as favoured by the leading biotech country, the United States, towards a more process-based approach as practiced in the EU. During the drafting of these new regulations, the Chinese authorities paid close attention to policy developments in Europe, Japan and South Korea, as well as the outcome of the Cartagena Protocol negotiations (interview with MOA representative, 23 August 2004). The impression of a shift towards a more restrictive policy was further deepened when in February 2002 MOA issued a ban on foreign investment in the biotech seeds business. This move seemed to play into the hands of those critics who had argued that China's new-found emphasis on the safety of GMOs was motivated more by protectionism than environmental protection (Newell, 2003b: 30; interview with Chinese Academy of Sciences representative, 22 August 2004; interview with industry representative, 24 August 2004).

Despite the evident strengthening of the biosafety provisions, however, the new system for biosafety management is still criticised by environmentalists for failing to reduce the central role played by pro-biotech scientists and regulators in the approval process, particularly in the powerful Biosafety Committee (Keeley, 2003c), and for maintaining the Agriculture Ministry's central role in the regulatory process without giving greater authority to the State Environment Protection Agency (SEPA), China's equivalent to an Environment Ministry. SEPA remains marginalised in the regulatory framework but has recently sought to acquire a greater role through its involvement in the current drafting of the first fully fledged national biosafety law under the direction of MOST (interview with SEPA representative, 17 August 2004).

As this brief historical overview suggests, China's biotech policy has undergone an important transformation since the mid-1990s. This brings us to the question 'how can we explain this policy shift?' What factors account for the strengthening of environmental concerns in GMO risk assessment? Conventional explanations of the 'greening of the state' as a result of pressure from social movements – as is evident in industrialised countries (Dryzek et al., 2003) – do not work in this case. Government insiders and representatives of civil society alike acknowledge that there has been no significant public debate, let alone political campaign, that might have sparked a governmental rethink on GMOs (interview with civil society representative, 22 August 2004; interview with SEPA representative, 17 August 2004). Much of the biosafety debate has been conducted in elite scientific and governmental circles, and the policy change described above has been initiated from within the state. Yet, as this chapter argues, debates within the Chinese state have themselves been subject to international influences and cannot be understood without considering the international context. In the following, the two most important external factors, international socialisation and economic globalization, will be examined in greater detail.

International socialisation and participation in global environmental governance

China's opening over the last three decades represents one of the most profound changes in the country's foreign policy and in international relations more generally. Since the People's Republic of China (PRC) was recognised in 1971 as the sole representative of China at the United Nations, and especially since Mao's death in 1976, China has joined a range of international organisations and committed itself to adopting international norms and rules (Kent, 2002). The Chinese leadership after Mao saw membership in international organisations as both a requisite and a boost for achieving great power status. Yet integration into international society has brought with it new threats to national autonomy, as China has had to adopt certain international norms and rules that stand in contrast to its domestic political constitution. The country is also exposing itself to greater international scrutiny of its record in implementing international obligations. China has persistently tried to strike a balance between international integration and national autonomy (Hughes, 1997). The difficulty in finding this balance is most prominently illustrated by China's accession to the World Trade Organization (WTO) in 2001, which became one of the most hotly debated foreign policy issues in China (Fewsmith, 2001a).

The rise of domestic environmentalism

A notable consequence of China's international socialisation has been the growing recognition of environmental concerns on the domestic political agenda (Economy, 2004). Having traditionally viewed environmental issues as a 'luxury' that should not threaten the country's economic aspirations (Sullivan, 1995: 243–4), Chinese leaders have over the last decade embraced the concept of sustainable development. The current leadership under Hu Jintao has repeatedly stated that sustainability is a guiding principle of its economic policy and stressed its desire to put economic growth on an ecologically sustainable footing (Xinhua, 2002).

While the rise of an environmental agenda was to some extent precipitated by domestic developments, domestic factors alone cannot explain it. To be sure, recent high-profile environmental disasters and disputes have played a role: local opposition to the building of the Three Gorges Dam highlighted the social and ecological damage caused by high-profile infrastructure projects; the massive flooding in the Yangtze River and Songhua River valleys in 1998 led to the government's public recognition of the problem of deforestation; and urban air pollution has become a major source of additional public health costs (Economy, 2004: ch. 3; Sullivan, 2005: 245–6). Yet, the growth of China's environmental movement (Cooper, 2006) remains a fragmented and, with the exception of the campaign against the Three Gorges Dam, localised phenomenon (Ho, 2001: 897–900). Environmental organisations, although having seen rapid growth since the mid-1990s, continue to operate under restrictions imposed by the central government (Yang, 2005: 51; Schwartz, 2004).

Given the constraints on domestic environmentalism, the international context has therefore been of particular importance to the rise of green issues on China's policy agenda. In the early 1990s, scholars speculated that the growth of international environmental concern would put pressure on China to address ecological threats (Frieman, 1994: 191). This has been borne out in that the global system of environmental governance created since the 1970s provides an important normative and regulatory context for the adoption of national environmental laws and regulations in developing countries, and especially so in China. During the 1990s, China signed up to 25 multilateral environmental agreements and entered into a number of bilateral and multilateral projects for environmental cooperation (Chan, 2004: 71–2). China is a party to the Kyoto Protocol on climate change and the Convention on Biological Diversity (CBD), and is actively supporting the Montreal Protocol's phase-out plan for ozone-depleting substances. China's successful record of compliance with the Montreal Protocol has led observers to conclude that, where financial aid and capacity-building are provided, the country now plays a constructive role in international environmental protection (Chan, 2004; Zhao and Ortolano, 2003).

The emergence of the biosafety agenda in China during the 1990s has followed this pattern of externally induced environmentalism. It provides a clear example of what Economy and Schreurs (1997) have described as the internationalisation of environmental politics. International concerns about the risks associated with GMO releases and efforts to deal with them preceded similar concerns and regulatory developments in China.

The international context: scientific exchange, biosafety negotiations and capacity-building

The internationalisation of biosafety policy took on several dimensions: international scientific cooperation; participation in UN-sponsored biosafety talks; and biosafety capacity-building activities funded by international donor agencies. Links between the Chinese scientific community and the international GMO debate played a key role in strengthening the domestic biosafety agenda in the late 1990s (interview with SEPA representative, 17 August 2004). Scientists have been privileged actors in this process, not only because of their authoritative claims to policy-relevant knowledge but also because of the legitimacy they possess vis-à-vis a political elite that has staked its future on the promise of technological progress and economic growth. Moreover, unlike environmental campaigners, scientists can express legitimate ecological concerns without being suspected of politically subversive intentions.

Given that an international biosafety debate had long been underway since the 1980s (Zedan, 2002) before China came to consider the need for domestic regulation, Chinese biotechnology experts were able to tap into a rapidly growing stream of international biosafety research and apply established methodologies of environmental risk assessment to the Chinese context. By the 1990s, Chinese biologists were well connected with international scientific communities, as a consequence of the gradual opening to international collaboration that was started as part of Deng's modernisation strategy. These international links were critical in stimulating domestic interest in biosafety. In a sense, they provided a vehicle for transmitting information about GMO risks and biosafety concerns, first into domestic scientific debates, and then into the wider regulatory and political debate (interview with Chinese Academy of Sciences representative, 18 August 2004).

The role played by Chinese scientists in the process of information transfer and diffusion of environmental concern is reminiscent of the contribution that scientists make in other areas of international environmental protection. As Haas (1995) argues, transnational expert groups, or 'epistemic communities', may seek to influence scientific and policy discourses within states with a view to promoting international environmental cooperation. Their role extends to raising awareness and creating concern, informing scientific and policy debates, and strengthening the hand of those forces within the state bureaucracy or political elite that are in favour of taking environmental action. They perform an important, yet all too often ignored, function in internationalising domestic environmental policy and transmitting international concern into domestic contexts. That Chinese scientists can play such a role has been noted before, in the context of the debate on ozone layer depletion in the 1980s (Zhao and Ortolano, 2003) and on climate change in the 1990s (Economy, 1997).

The emerging debate on scientific biosafety concerns was further promoted by China's participation in the international negotiations on a biosafety treaty conducted under the auspices of the CBD between 1996 and 2000. This international process brought together international scientists, regulatory experts, environmental NGOs and industry groups in an effort to establish the parameters and modalities for regulating transboundary movements of GMOs and culminated in the adoption of the Cartagena Protocol on Biosafety (Bail, Falkner and Marquard, 2002). Having been one of the first parties to ratify the CBD in 1993, China participated in these talks from the outset. Its delegation of scientific and regulatory experts was led by the State Environmental Protection Agency (SEPA), which had hitherto played only a marginal role in domestic GMO governance. China joined what came to be known as the Like-Minded Group of developing countries, the key *demandeur* for a strong biosafety regime, but maintained a relatively low profile within that group, particularly when compared to the more active representatives of Ethiopia and India. At critical junctures throughout the negotiations, China showed willingness to reach a compromise in order to secure an agreement (Lijie, 2002), but mostly followed the group's position.

China's involvement with the biosafety talks had an important impact on its domestic agenda: it exposed China's scientific and political elite to the full range of biosafety concerns and risk regulation approaches being debated worldwide and helped to boost the standing of environmental experts domestically. The early phase of the negotiations, from 1996 to 1998, was concerned primarily with defining the key issues and concepts of international biosafety regulation and produced a long 'wish list' of items that delegates wanted to be included in a future biosafety treaty (Falkner, 2002a: 7–14). For the scientists involved, it provided a unique opportunity for international dialogue and the exchange of information on biosafety-related research that had been conducted around the world (Gaugitsch, 2002). Chinese representatives taking part in the biosafety negotiations thus became a key element in the link between the international and domestic debate on GMO safety. As Wang (2004: 902) writes, the biosafety negotiations during the late 1990s 'intensified the concerns and awareness of potential adverse effects of GMOs on environment and health.'

Participation in the Cartagena Protocol negotiations also gave rise to an institutional mechanism for transmitting biosafety concerns and norms into the domestic context. In order to support the regime-building effort and implementation of the protocol, the international community provided assistance for capacity-building to developing countries and economies in transition. China became one of the largest recipients of biosafety capacity-building aid from the late 1990s onwards. These projects, which were funded mainly by bilateral and multilateral donor agencies, aimed at establishing the scientific, administrative and regulatory capacity needed for carrying out GMO risk assessment and management. One of the key elements in China's capacity-building effort was the creation of a national framework for biosafety, which has been running from the late 1990s onwards and was sponsored by the Global Environment Facility (GEF) and UNEP. The country's environmental agency SEPA took this opportunity to challenge the dominance of other ministries, especially MOA, in domestic GMO regulation and was able to establish its lead role in the framework drafting process. The impact of the national framework, however, was of a more indirect nature. The initiative established a process that did not directly feed into the evolving domestic regulatory process, and SEPA's efforts eventually to replace MOA as the lead regulatory agency have not come to fruition. Still, the drafting exercise provided for yet another opportunity to connect the regulatory debate in China more closely with developments abroad, and it galvanised scientific and political interest in biosafety matters (interview with SEPA representative, 17 August 2004).

Thus, the consequences of these forms of international concern and norm transmission were threefold. First, they helped to raise awareness in China of the environmental and health risks arising from genetic engineering in agriculture and legitimised a biosafety discourse that challenged the predominant focus on promoting technological advances in biotechnology. Second, the growing attention paid to biosafety issues and participation in the biosafety negotiations have allowed SEPA to raise its own profile domestically, albeit with mixed results. Third, despite SEPA's failure to change the institutional basis for risk assessment, greater awareness of GMO risks has led to a strengthening of biosafety regulations. The State Council's new 2001 Regulation on the Safety Administration of Agricultural GMOs was followed in 2002 with three detailed implementation regulations issued by MOA, providing for the first time a comprehensive risk assessment of both domestic GMO applications and imports of GMOs. These new regulations updated and deepened the regulatory system put in place in the mid-1990s and reinforced the emerging shift towards a more precautionary approach in commercialising GM crops.

Economic globalization and environmental 'trading up'

Besides participation in international scientific debates and environmental regime-building, China's growing integration into the international economy has provided a further, potent, stimulus for the growth of a domestic environmental policy agenda. Intensified trade links and greater exposure to foreign investment have forced China to consider the implications that environmental issues may have for its competitive position globally. The case of GMO politics is an important example of how greater economic integration can have a 'trading up' effect on environmental policy.

China's economic globalization

The link between globalization and domestic environmental policy has been a hotly contested issue in academic and policy circles. Many contend that global economic competition has a detrimental effect on environmental quality. In their view, increases in trade and investment are seen as spreading an environmentally unsustainable pattern of production and consumption and add to existing pressures on the Earth's finite resources (Goldsmith and Mander, 2001). But economic globalization can also act as a force for promoting environmental agendas and greener business practices. As Vogel (1995) and others (Princen, 2004) argue, an increase in economic interdependence is not necessarily linked with a weakening of national environmental standards. On the contrary, economies that are among the most internationally integrated also have some of the highest environmental standards. Countries with stricter environmental regulations can create export opportunities for environmental protection services and goods; and increased regulation in import markets may force export firms to improve the environmental quality of their goods. Thus, greater economic integration may provide a framework for 'trading up' instead of 'trading down' environmental standards (Vogel, 1995), and countries that are subject to economic globalization may end up increasing regulatory standards rather than engage in a 'race to the bottom'. The 'trading up' scenario has so far been documented primarily in industrialised countries, but recent studies suggest that developing countries may also experience upward regulatory pressures resulting from globalization (Garcia-Johnson, 2000; Vogel, 1997: 557).

In recent years, China has taken important steps towards greater integration into the international economy. The gradual opening of China's state-controlled economy has been an integral element of the country's reform process. The most significant step on this path occurred in December 2001 when China became a member of the World Trade Organization (WTO). In doing so, China committed to adopting a whole range of international trade norms and obligations based on the WTO's enduring objective of liberalising international trade. The preparations for WTO entry further accelerated China's deepening engagement with the global economic order: in 2003, China overtook the United States as the world's leading destination for FDI, with annual FDI flows amounting to $53.5 billion (UNCTAD, 2004); and having recorded a year-on-year growth in trade during the 1990s, China's total merchandise trade in 2003 nearly matched that of Japan (WTO, 2004: 5).

Globalization and global market integration are less pronounced in agriculture than in other sectors, such as manufacturing, but agricultural trade has steadily grown at an average rate of 6 percent during the 1980s and 1990s (Huang and Rozelle, 2003: 116). Since the WTO accession agreements were signed, China has stepped up its efforts to open its vast agricultural market to foreign imports. WTO entry has further strengthened the structural shifts in agriculture away from a model of self-sufficient production (for example, emphasising grain) to higher value and less land-intensive production (for example, horticulture, livestock). While globalization of the country's farm sector is still in its infancy, the effects of greater international competition are being felt particularly in those agricultural sectors, such as soybeans, where trade barriers have been lowered and imports have risen dramatically (ibid, 119).

'Trading up' in agri-biotechnology

The 'trading up' effect on agricultural biotechnology was driven by the fear that should China go ahead with commercialising GM crops it would end up losing market access to countries with GMO import restrictions. The first time that this dilemma occurred was in the case of GM tobacco, China's first commercially grown GM crop that was introduced in the late 1980s, long before other countries began the commercial planting of GM crops. The virus-resistant GM tobacco variety promised higher yields to producers but fell out of favour with international buyers, particularly – and ironically – the United States, which was concerned about negative consumer reactions to the use of GM tobacco. China responded by officially ending the planting of GM tobacco in an attempt to placate and retain its tobacco export markets, although the precise extent to which farmers have phased out GM varieties remains unknown (Paarlberg, 2001: 128–9). The experience with GM tobacco did not in itself put an end to China's biotechnology programme, but provided a first glimpse of the repercussions that the domestic introduction of GM crops might have on export markets (interview with Chinese Academy of Sciences representative, 18 August 2004).

GM cotton became the second major transgenic crop variety after tobacco to receive regulatory approval in China. In 1997, the first insect-resistant GM cotton varieties passed the regulatory hurdles and were introduced in four provinces (Hebei, Henan, Shanxi and Shandong) around the Yellow River, in an area severely affected by bollworm in the early 1990s (Keeley, 2003c: 10). For the first time, China also allowed a foreign company, Monsanto, to sell GM seeds. The initial success of the Bt cotton varieties was evident from the rapid adoption rates: by 2001, GM cotton was being grown on more than 2 million hectares and accounted for 45 percent of China's cotton area, with nearly 5 million Chinese farmers planting the GM variety (Huang and Wang, 2003: 11). But because cotton has been grown primarily for domestic consumption and does not enter the food chain, biosafety concerns surrounding GM varieties (for an overview, see Keeley, 2003c: 18–27) were not affected by international trade concerns. Thus, in contrast to the tobacco experience, agricultural globalization did not have a noticeable 'trading up' effect on GM cotton.

The threat of exclusion from export markets resurfaced, however, in other GM crop developments that affected food production and China's international trade. In the late 1990s, some of the world's biggest agricultural import markets began imposing labelling requirements and restrictions on GM food shipments. The EU led the way with its *de facto* moratorium on GM imports of late 1998 (partially lifted in 2004). The temporary ban on new GM crops was essentially an effort to restore confidence in the EU's system for food regulation that had been damaged by a string of food scares culminating in the outbreak of 'mad cow disease' (BSE). The first GM foods to be sold in Europe had been met with growing consumer concerns and were described in the media as 'Frankenstein foods'. Although the EU's regulatory system was designed to ensure that health and environmental concerns were taken into account in the GMO approval process, some EU member states felt it necessary to impose a ban on further approvals to avert an even greater crisis in public confidence. In 2003, the EU further strengthened its regulations by introducing a comprehensive system of labelling and traceability for all GM food content, whether present in the end product or during the production process, thus giving the consumer the right to choose between GM and non-GM food (Brack, Falkner and Goll, 2003). This meant, however, that continued public hostility to GM crops and food in Europe could close off the European export markets for GM products grown outside the European continent – a development that was closely watched in China.

The EU was not alone in taking precautionary action against GM food. Japan and Korea, two of China's most important agricultural importers in Asia, also introduced restrictions on GM imports and established their own labelling systems for GM content in food. Both countries are important import markets for Chinese soy-based products, such as tofu and soybean sauce. These developments amounted to a serious challenge to China's desire to promote agricultural biotechnology and mobilised domestic export interests to lobby against the proposed introduction of GM soybeans in domestic production (interview with Chinese Academy of Sciences representative, 22 August 2005).

The threat of market exclusion caused by rising anti-GM sentiment in Europe and elsewhere became all too real when in 2000 the EU temporarily halted imports of Chinese soy sauce after British inspection authorities detected GM content. Although China did not allow the commercial growing of GM soybeans domestically, it had started to import GM soybeans from the United States in order to meet growing soybean demand in the production of animal feed and food products that were also destined for export. GM soybean varieties were being developed and tested by Chinese research laboratories at that time, but a decision on authorising domestic growing of GM soybean has since been put on hold. Chen Zhangliang, a leading biotechnology expert and the then vice president of Peking University, summed up the growing concerns over the loss of export markets:

> We fear bans on the export of our products. This is a big controversy in Europe . . . Some have proposed special areas set aside for export goods, where GM crops would not be used. But this is hard to enforce since everyone wants to export. The government is very cautious. (O'Neill, 2001)

Externally induced market pressures thus helped to shift Chinese policy towards a more precautionary stance. To protect its soybean export markets, China postponed a decision on whether to allow domestic cultivation of GM soybeans. It also established a segregation system to exclude GM soybean imports from any domestic use other than in crushed form as animal feed, the effectiveness of which, however, remains contested (interview with industry representative, 24 August 2004; interview with civil society representative, 22 August 2004).

The introduction of more stringent biosafety rules has also led to disruption in China's farm imports and has frustrated efforts by foreign firms to develop a GM crop base in China. Among the new trade-related regulations introduced in March 2002 was a requirement for foreign GM crop shipments to receive safety certification before they could be imported. The rules caught China's biggest trading partners by surprise and led to a fierce diplomatic spat with the United States over the disruption they caused in US soybean shipments (Rugaber, 2002). Introduced only 4 months after China's WTO entry, the new certifications system was interpreted in Washington as 'back-door' protectionism aimed at manipulating the burgeoning trade in soybeans flowing into China, the single-largest export market for US soybean producers (interview with US trade official, 25 August 2004). US trade representatives complained about the uncertain nature of the new biosafety rules, which failed to give clear guidance to traders on the documentation requirements and allowed Chinese authorities to delay a decision for up to 270 days, a time frame China adopted from the Cartagena Protocol's procedures for risk assessment. A fall in soybean imports from the US was only reversed after the country bowed to US pressure, including from President George W. Bush, and produced interim safety certificates for US imports before issuing formal 3-year certificates in February 2004 (*China Daily*, 2004; interview with Ministry of Commerce official, 24 August 2004).

In sum, greater exposure to international economic competition has helped to strengthen those within China's scientific and political elite who have argued for a more precautionary approach to commercialising GM crops. Commercial, and to some extent protectionist, interests have served to reinforce environmental arguments in favour of a 'go-slow' approach to commercialising GMOs. This may change, of course, should international trade restrictions on GM food be lifted in the future. The persistence of the trading up effect also depends on the continued existence of sufficiently strong farm export interests in China that perceive the introduction of GM crops as a commercial risk, as has been the case with soybeans and corn. As Anderson and Yao (2003) argue on the basis of an economic analysis of China's trade sensitivity in agriculture, these export interests are likely to persist in the foreseeable future. In their view, China has '…a vested interest in ensuring that the GM debate abroad does not lead to excessive denials of market access for GM products' (Anderson and Yao, 2003: 169).

Conclusion: globalization, internationalisation and environmental policy change in China

This chapter has examined the recent shift in China's biotechnology policy towards comprehensive risk assessment and a more precautionary approach to commercialising GMOs. It has argued that in order to fully understand the sources of this policy change we need to consider the impact of international political and economic factors on domestic environmental policy-making. This is in line with recent arguments in the debate on

China's internationalisation and domestic reform. Chinese leaders increasingly recognise the need to adjust to international norms (Moore, 2000), a trend that is being reinforced by China's integration into the global economy (Zweig, 2002).

Two dynamics have been identified as the international factors driving this process: international socialisation, as seen in the growing integration of China into international environmental governance and transnational scientific networks; and economic globalization, particularly the environmental 'trading up' effect brought about by greater exposure to international agricultural trade. Both these dynamics have worked together in this case to push environmental concerns higher up the political agenda in China, against the background of comparatively weak domestic pressure from either consumers or environmental campaign groups.

The case of GM food regulation provides an intriguing example of the ways in which the growth of international linkages and transnational actor networks can bring about domestic environmental policy change. This case is all the more significant as China has traditionally placed a high value on preserving its national sovereignty. Even in the current era of economic liberalisation and international integration that Deng's reform policy initiated, China has been adamant in its defence of policy autonomy vis-à-vis external influences. It has sought to curtail the growth of a domestic environmental movement and to monitor transnational links with international environmental groups. This is not to say, however, that China has experienced a loss of autonomy in the field of biosafety policy. Far from it, the Chinese state continues to exercise a high degree of control over both the development of biotechnology and regulatory policies in biosafety. The internationalisation of biosafety policy has led to a reconfiguration of the power balance between pro-biotech and pro-environmental advocates within the core state. Although civil society groups are slowly making their mark in this struggle, the decision-making process over the future of biotechnology remains firmly in the hands of the state. Moreover, the rise of the biosafety agenda has given the state a different means of controlling the direction of the biotechnological revolution, particularly with regard to the growing influence of foreign biotech firms. It has also added a new regulatory barrier to international farm trade that, as some of China's main trade partners allege, has gained in popularity as other trade barriers have been eliminated as a consequence of WTO accession. The shift towards greater precaution in GMO regulation, therefore, reflects both greater environmental sensitivity *and* the strategic use of safety concerns for protectionist purposes. In a sense, the biosafety agenda satisfies both the environmental constituency and those state elites that seek to control foreign investment and competition, while integrating the country ever further into the global economy.

The Chinese experience will be familiar to many other developing countries that are receptive to international biosafety concerns and are keen to preserve agricultural export markets in regions with GMO import restrictions. Many African and Southeast Asian countries have found themselves in such a position. Other countries, however, will face different sets of external influences. Mexico and Brazil, for example, have recently relaxed their restrictive GMO policies while liberalising trade with the United States, the world's leading biotech country. Thus, the nature of integration into the international political economy will shape the direction in which external factors drive domestic policy change in the developing world.

It is nevertheless important to recognise the limitations of the internationalisation effect on China's environmental policy. As in the case of climate change policy, where a 'greening' of policy has been thwarted by pro-developmental interests within the core state (Economy, 2004: 183), the move towards precaution in agri-biotechnology is resisted by powerful interests within the scientific community and government. Environmental risk assessment of GMOs remains contested and is far from being fully established. Indeed, the Chinese government remains committed to utilising the potential of genetic engineering and continues to support research and development of new GM products. Over the last few years, a group of scientists have intensified their efforts to convince the government of the need to authorise new GM crops, especially GM rice, in which China has developed a competitive edge. Concerns over biosafety and the impact of future rice exports have so far caused the government to delay a decision on GM rice, but the balance of concerns is tipped by some observers to shift in favour of commercialisation in the near future (interview with Chinese Academy of Sciences representative, 22 August 2004).

As in other instances of political internationalisation and economic globalization, domestic institutions and interest group politics play an important role in shaping the way in which international factors impact on the domestic level. Very few, if any, countries simply adopt international norms and obligations, and most countries are able to ameliorate, at least to some extent, the effects of global economic integration. Through processes of transformation into national law, domestic implementation, legal interpretation and even international re-negotiation, they respond to international commitments in an often creative manner, thus preserving a significant degree of national autonomy in the transmission of international commitments. The case of GMO policy in China supports this view. The different international forces operating in the field of biotechnology have been utilised by competing interests within the Chinese state and the biotech sector to support their own agenda. While economic and political globalization have created a more open process of deciding the future of biotechnology in China, domestic actors within, or associated with, the core state, will largely determine the direction of this process.

What has changed under conditions of globalization is that these debates over the use of biotechnology in agriculture no longer take place solely within the confines of a scientific–industrial biotech complex that has been the driving force behind the country's advances in agri-biotechnology. They are now conducted with the legitimate participation of ecological researchers and biosafety regulators who are keen to ensure that environmental risks receive due attention. Moreover, civil society is slowly making its way into this debate, with Greenpeace leading a careful and targeted campaign to disseminate scientific research and inform mainly urban consumers of the risks involved in GM food. China's global economic and political integration has thus set the scene for environmental policy change in an important area.

Notes

1. Research towards this chapter was supported by a grant from the John D. and Catherine T. MacArthur Foundation. The author expresses his gratitude to all Chinese government officials, researchers and representatives of NGOs and companies who were willing to be interviewed in August 2004. Thanks are also due to Jennifer Clapp, Manfred Elsig and Aarti Gupta for helpful comments on an earlier version of this chapter, and to Chen Zhimin, Chris Hughes, Brendan Smith, Song Xinning and Xu Ang for facilitating research in China. The usual disclaimers apply.
2. In September 2005, China approved the commercialisation of a new Bt cotton variety, but new GM crops that would enter the human food chain have so far failed to receive regulatory approval (*China Daily*, 2005).

Part V Biotechnology Regulation, the WTO and International Environmental Law

11
The Biosafety Protocol and the WTO: Concert or Conflict?

Grant E. Isaac and William A. Kerr

Introduction

As the international research, development and commercialization of genetically modified (GM) agricultural crops increases, so too does the need for an internationally accepted regulatory framework capable of balancing the potential benefits of products derived from these crops with their potential risks. Such a framework is difficult to establish, however, because regulatory decision-making structures are inextricably linked to domestic political economy factors. As these factors can differ significantly from jurisdiction to jurisdiction, so too can regulatory structures. As a result, GM crops approved for market access in one domestic market may not be approved in another resulting in trade tensions that can become trade disputes. On 13 May 2003, the United States (US) along with Argentina, Canada and Egypt requested formal World Trade Organization (WTO) consultations on the European Union (EU) moratorium on the approval of genetically modified organisms (GMOs). In a preliminary ruling released on 7 February 2006, the WTO's Dispute Settlement Body agreed with the complainants that the EU moratorium on the approval of GMOs was incompliant with the international trade rules. This preliminary decision was confirmed on 12 May 2006.[1]

Despite the decision, the GMO trade dispute remains an important area of study because it represents a complex intermingling of crucial political forces. It is about the ability of sovereign states to independently regulate technological innovations as the World Trade Organization's disciplines increasingly reach into domestic policy competence. The US, along with Canada, supports a particular 'North American' regulatory approach to products of modern biotechnology that is different from a much more complex regulatory approach supported in the EU. In other words, transatlantic regulatory regionalism exists and the US and Canada would like the WTO to determine whether or not the EU regulatory approach is compliant with current international trade obligations. This, of course, takes the WTO out of its traditional focus on border measures and into the very controversial area of adjudicating on the appropriateness of domestic regulations.[2]

The GMO dispute also occurs within the context of both the strained relations between the US and 'Old Europe' and the stalled momentum at the Doha Development Round of WTO negotiations over agricultural policy issues. With respect to the former, the end of the Cold War marked a decoupling of high politics (security and stability) from low politics (trade relations) allowing many contentious trade issues to emerge on the transatlantic agenda such as conflicts over corporate taxation standards, bananas and hormone-treated beef. These events were often viewed as minor irritants of low politics among good friends. However, the recent war in Iraq pitting the high politics of US foreign and security policy against that of 'Old Europe' led by France and Germany marked what may be an unprecedented low in the transatlantic relationship. With respect to the Doha Development Round, this WTO round of negotiations on trade liberalization is essentially on hold until the US and the EU prove to the less-developed countries that they are serious about global welfare gains by liberalizing their well-protected agricultural sectors. This means both sides taking on very powerful domestic interests. As both argue that their agricultural policies are in place to protect their domestic producers from the distortions created by the subsidies and tariffs put in place by the other, any movement on agricultural trade liberalization will require a simultaneous ratcheting down of policies (Gaisford and Kerr, 2003). Achieving this requires a willingness for cooperation and compromise which appears notably absent from the current transatlantic relationship.

The GMO dispute also illustrates the uneasy relationship between trade and environmental protection measures. While the EU is the explicit target of the trade action, an implicit target is the Cartagena Protocol on Biosafety (also known as the Biosafety Protocol). The Biosafety Protocol is a multilateral environmental agreement (MEA) that specifies rules for the transboundary movement (trade) of products of modern biotechnology in order to protect biodiversity. The Biosafety Protocol is of concern to the US and its co-complainants because it basically *multilateralizes* the EU regulatory approach by potentially allowing other countries to use the Protocol to justify adopting EU-style market access rules. Therefore, the WTO challenge to the EU moratorium on GMOs can also be viewed as a challenge to the Protocol and, as such, a signal to all other countries who might be tempted to use the protocol to ban GMOs. Similar to the interface of trade agreements and domestic regulations, the interface of trade and environmental agreements takes the WTO into very controversial territory.

The GMO trade dispute has the potential to lead to a further deterioration in transatlantic economic and political relations and to dramatically undermine not only the Doha Development Round of WTO negotiations but also the very legitimacy of the WTO. Consider first the recent outcome where the WTO panel ruled in favour of the US and Canada. This ruling will be portrayed as both a decision *for* biotechnology (and the large multinational companies that have championed its commercial development) *against* the human, animal and environmental health and safety regulations of the EU and also as a decision against the·Biosafety Protocol and the protection of biodiversity from products of modern biotechnology. Beyond just the international trade of products of modern biotechnology, such a ruling represents another decision against the EU – like the hormone-treated beef case[3] – and potentially decreases the willingness of the EU to undertake the real reform of its Common Agricultural Policy required to re-invigorate the Doha Round. Such a decision could also amplify concerns about the WTO's legitimacy embodying the fears of the WTO's harshest critics that the WTO is an unaccountable international force that reaches deep into domestic policy competence constraining health and safety policy options. The decision against the Biosafety Protocol would also embody the more general criticisms of environmentalists that free trade is at the expense of the environment.[4] Simply put, transatlantic relations, the Doha Round and the legitimacy of the WTO are at stake.

As this introduction indicates, establishing an internationally accepted regulatory framework for GMOs has been highly controversial. This chapter provides an examination of this issue through a comparative assessment of the regulatory frameworks supported under the Biosafety Protocol and the World Trade Organization with an aim of identifying areas of concert and conflict. In the next section, the context for this examination is set: the political economy of regulations and science, regulatory systems analysis, regulatory regionalism and international policy coordination. This is followed by the comparative assessment and in the last section implications and conclusions are drawn.

Analytical context

This examination of the degree of concert and conflict between two international regulatory frameworks can be understood within the context of two interrelated academic discourses: *regulatory systems analysis* and *international policy coordination*.

Regulatory systems analysis

The study of the origins, principles and functions of regulatory systems reveals two general insights. The first is that they are inherently bottom-up in nature. That is, they are a function of political and social responses to domestic experiences with regulatory success and failure (Black, 2002; Beck, 1999). Second, regulatory systems are moving away from the state-centered construction to systems that include a mix of public and private actors involved in all stages of regulatory development from rule-making, to monitoring for compliance and to enforcement (Scott, 2002). Alternative systems can have private governance mechanisms, or they can have management-based performance structures or can involve co-regulation where both public and private regulators are engaged or self-regulation which engages only private regulators (Scott, 2002; Gunningham, 1995). Together, these two features imply that as jurisdictions differ with respect to experiences as well as with respect to the mix of governance mechanisms, performance measurements and public and private actors, the regulatory systems that emerge will be different: the so-called asymmetrical regulatory systems.

The asymmetry complicates comparative assessments of regulatory systems because i can corrupt the comparability of systems, where typical evaluative criteria include regulatory legitimacy, accountability and effectiveness (Black, 2002; Picciotto, 2002). To remedy this, a regulatory systems analysis could be employed (Isaac, 2002). Such an analysis goes beyond a comparative catalogue of rules and attempts to identify the underlying regulatory principles upon which the rules are based. These principles include the bottom-up belief systems and norms that prevail in a particular jurisdiction.

There are three key advantages of a regulatory systems analysis. First, it can identify regulatory path dependency and, consequently, allows for the dynamic study of regulatory resilience (Hartley and Skogstad, 2005; Skogstad and Moore, 2003). Second, it allows for a comparative analysis of the operationalization of the regulatory approaches and not just an examination of the rules. As a result, it creates a context for capturing the reality of the diversity of rules-based approaches. The third advantage of a regulatory systems analysis deals with the opportunities to link policy domains; in this case regulatory, trade and environmental policy. The source – and scope – of the dispute can be more clearly identified: is it a different rule or a more complex divergence arising from differences in the principles underpinning regulatory rules?

International policy coordination

In a broad sense, *international policy coordination* has the objective of solving policy (mis)alignment issues between sovereign nations – often with very different political economy situations – through bilateral, plurilateral or multilateral policy rapprochement negotiations. For example, monetary policy coordination represents an attempt to control the international spillover effects that foreign policies can have on domestic inflation and employment (Persson and Tabellini, 2000). International trade policy coordination represents an attempt to solve multinational market failures preventing the efficient allocation of resources that arise when nations do not engage in economic activities consistent with their comparative advantage (Gaisford and Kerr, 2001). Similarly, international environmental policy coordination represents an attempt to solve problems of environmental degradation that arise when nations acting in their own best (economic) interest fail to act in the best interest of global biodiversity (Helm, 2000; Killinger, 2000).

International policy coordination is rarely easy. For instance, multilateral efforts to liberalize international trade, such as the Uruguay Round and the current Doha Development Round, often take much longer than initially expected and often involve several significant crises that threaten the entire round. Even at a regional or plurilateral level, policy coordination is difficult as the European Union's efforts to establish the Single European Market or the continuing softwood lumber dispute between Canada and the US demonstrate. The challenge for international policy coordination is that the objective of maximizing global gains is sometimes not consistent with that of maximizing domestic gains. A national government, elected by domestic – not international – constituents, may simply lack the political will to engage in activities that increase global welfare at the expense of domestic welfare. In addition, if one nation faces this dilemma, then others do as well, creating a market failure of collective action problem (Olson, 1965).

Successful multilateral efforts in international policy coordination may result in the creation of multilateral regimes paradigms. For example, the creation of the World Trade Organization (WTO) in 1995 from the Uruguay Round of General Agreement on Tariffs and Trade (GATT) negotiations represents the international trade paradigm codifying the rights and obligations of nation-states who wish to be WTO members. Similarly, the creation of the Convention on Biological Diversity (CBD) from the 1992 UN Conference on Environment and Development represents the international biodiversity paradigm codifying the commitments to the protection of biodiversity made by the CBD's ratifying nations. Typically, there is overlap between such independent paradigms where sometimes in the nexus there is policy concert, other times there is conflict. Understanding the factors that account for concert and conflict is crucial in ensuring that any benefits from policy coordination that may be achieved in one paradigm are not eroded through conflicts with another paradigm. While there is a significant body of literature on how these multilateral paradigms emerge (Gilpin, 2001), there is a limited amount of literature on what happens when these multilateral paradigms overlap (Isaac et al., 2002).

Previous *international policy coordination* research (Isaac, 2003) employing an institutional analysis methodology across the 32 most trade-distorting MEAs has revealed two factors common to a harmonious trade–MEA relationship. First, when the environmental issue tackled by the MEA is narrow, there seems to be an increased likelihood that nations are willing to abide by the environmental protection measures even if they may, in fact, hinder international trade because they do not spill beyond this narrow issue. The trade in endangered species, ozone-depleting substances and hazardous wastes all fit this criterion of being narrow environmental issues. It appears that while the first seems to be a necessary condition it is not a sufficient condition for a harmonious trade–environment relationship. The second important factor is the presence of a transatlantic agreement on the issue. That is, if the US and the EU can agree that the environmental measure deserves attention and that the regulatory response outlined in the MEA is acceptable (that is, there is consistency between the US and the EU regulatory approaches), then the likelihood of conflict with the international trading regime appears to decrease. Again, protection of endangered species and the ozone as well as domestic biodiversity from hazardous wastes are policy goals shared with virtually equal fervour on both sides of the Atlantic.

Comparative analysis

In this section, an institutional analysis methodology – drawn from both the regulatory systems analysis and the international policy coordination literature – is adopted to assess the degree of concert or conflict between these two multilateral paradigms. An institutional analysis methodology is appropriate for the comparative analysis of the similarities and differences among institutions and paradigms. Accordingly, comparators are identified typically consisting of, but not limited to (1) origins and nature of the institution, (2) mandate/scope of the institution, (3) regulatory principles as well as (4) regulatory decision-making approaches.

Origins and nature

The World Trade Organization is an international organization created during the Uruguay Round of negotiations on the GATT and that came into force on 1 January 1995. Members to the organization must be nation-states.[5] The WTO Secretariat – located in Geneva – is responsible for the administration of the various WTO Agreements such as the Agreement on Sanitary and Phytosanitary Measures (SPS Agreement) and the Agreement on Technical Barriers to Trade (TBT Agreement).

The Biosafety Protocol is an international treaty (and not an international organization). Its establishment was called for in Article 19(3) of the 1992 Convention on Biological Diversity (CBD). The Protocol was adopted by the Conference of the Parties (COP) to the CBD in January 2000 in Montreal, Canada, and entered into force on 11 September 2003 (after 50 signatory countries had ratified it).

Mandate

The WTO has a narrow mandate of trade liberalization for goods and services based on the principle of non-discrimination (PND) (Isaac, 2002). There are basically three concepts embedded in the PND. The first is the concept of *like* products whereby trade agreements do not focus on how a good (or service) is processed or on the production method used (PPM), but rather on the end-use attributes of the good (or service). A cotton shirt is *like* a cotton shirt regardless of whether the cotton was produced in an intensive or organic agricultural system. The second concept is that of *national treatment* whereby foreign goods or services must be treated the same in terms of market access rules as *like* domestic goods and services. The third concept is that of *most-favoured nation* whereby the favourable market access enjoyed by one particular foreign producer must be extended to all foreign producers of *like* products. Together, these three concepts combine within the PND to produce multilateral reciprocity. Moreover, they become the default principle whereby domestic measures are trade compliant if they pass the default test of non-discrimination. The benefit of this system is that it moves towards a rules-based system allowing for international trade to be a commercial function and not a government-to-government function.

Recognizing that the principle of non-discrimination cannot always apply, specific trade agreements such as the SPS Agreement outline the instances where a country can legitimately violate the PND (Isaac and Kerr, 2003a). For example, if Canada has a scientifically sound reason for banning a particular foreign product because of a risk to human, plant or animal health, that ban does not have to apply to all foreign and domestic producers of *like* products.[6] That is, any or all of the three concepts of non-discrimination – *like* products, *national treatment* and *most-favoured nation* – may be suspended by the importing country at its own discretion.

When the techniques and procedures of genetic modification produce crops with production traits (that is, without any output traits that would make the product distinguishable from non-GM crops) they are considered to be *like* products under the international trading regime. Therefore, due to the absence of scientific justifications for banning GM crops, the WTO – consistent with the PND – does not explicitly focus on market access for GM crops as distinct from non-GM crops (Isaac and Kerr, 2003a). In other words, the WTO supports a *product-based* approach to GM crops (Isaac, 2002).

The Biosafety Protocol is an MEA with a mandate to protect environmental biodiversity from the transboundary movement (that is, trade) of living products of modern biotechnology[7] with a scope that applies to:

> the transboundary movement, transit, handling and use of all living modified organisms that may have adverse effects on the conservation and sustainable use of biological diversity, taking also into account risks to human health.[8]

The Protocol supports a *process-based* approach whereby it is the use of modern biotechnology – regardless of the impact upon the end like product – that triggers regulatory oversight. More fundamentally, while the WTO's underlying regulatory principle is the principle of non-discrimination, underlying the Biosafety Protocol is the principle of advance informed agreement (PAIA). Modelled initially on the Basel Convention (Isaac, 2002), the Protocol essentially treats products of biotechnology in analogy to hazardous waste whereby the government of the importing country (Party of Import) must be notified by the government of the exporting country (Party of Export) of the intended transboundary movement of living products of biotechnology to allow the Party of Import to conduct its own risk analysis and determine the risk to domestic biodiversity prior to the shipment. Without a link to the international trading regime or to international scientific organizations, the Biosafety Protocol embraces a broader range of factors that Parties of Import may consider during market access decisions. These range from fairly standard approaches to risk assessment and to factors taking into account socio-economic impacts resulting from market access. That is, according the Protocol, Parties of Import may impose trade restrictions that potentially exceed those permissible under the WTO rules. Therefore, while the WTO aims at removing governments from the act of market access (a commercial activity), the Biosafety Protocol elevates the role of government making the transboundary movement of living modified organisms a government-to-government activity. Combining the process-based approach with the principle of advance informed agreement results in a highly precautionary protocol that treats products of modern biotechnology as potentially hazardous materials such that Parties of Import have sufficient room to make unilateral market access decisions while exporters and Parties of Export have virtually no recourse.

To elaborate on the differences between trade agreements and MEAs, consider the following example. Assume first that country A has recently put in place an environmental protection measure permitting organic agricultural production and banning intensive agricultural production practices. Next, consider two cotton shirts of which one is produced under organic standards of cotton production and the other is produced under intensive agricultural conditions. In country A, according to the environmental measure, only domestically produced organic cotton shirts would be permitted. Now consider two countries – B which produces organic cotton shirts and C which produces intensively produced cotton for shirts – who both wish to export to country A. The environmental position in country A would be to allow country B's exports, but deny country C's exports. However, the trade position is much different. According to the principle of *like products* B's and C's exports would be determined to be similar because their different PPMs do not create different end-use features; they are both cotton shirts. Hence, the trade position to allow both B's and C's exports into country A would clash with country A's PPM-based environmental objectives. Trade dispute rulings in both the tuna-dolphin and the shrimp-turtle cases at the WTO confirm these results.[9] In both cases, the United States attempted to impose trade restrictions on imports of tuna and shrimp products harvested with certain techniques that were incompliant with domestic PPM-based environmental objectives. The panel decisions essentially supported the position that if the tuna and shrimp harvesting techniques did not result in impacts upon the end-use characteristics of the subsequent food products then to target certain imports based upon US environmental preferences violated the like products concept of the principle of non-discrimination.

Regulatory principles

At first glance, it appears that these two international regulatory frameworks are very much in concert on at least three dimensions. First, they both support the use of the Risk Analysis Framework (RAF) in making market access decisions – WTO's SPS Agreement Article 2(2) and Biosafety Protocol's Article 15 and Annex III. The RAF was developed to deal with the regulation of advanced technology products (which were characterized by a large information gap between the producers of the innovation and the intended consumers) where the goal was to credibly inject science into public policy development (National Academy of Sciences, 1983).[10] The language of risk analysis is found in regulatory guidelines for the research, development and commercialization of advanced technology products – including GMOs – in many countries, including both the EU and the US, and in various multilateral agreements and treaties.[11] Second, they both support the use of the precautionary principle – WTO's SPS Agreement Article 5(7) and Biosafety Protocol's Preamble and within the Convention on Biological Diversity. Third, they both consider other legitimate factors beyond science in regulatory decision-making – WTO's SPS Agreement Article 5(3) and Biosafety Protocol's Article 26.

Yet upon closer inspection, it is revealed that these similarities are only superficial. In fact, while they share similar nomenclature, these frameworks differ considerably beginning with the fundamental basis for regulating technology in general. Two quite distinct regulatory trajectories have come to dominate GMO regulations – the scientific rationality trajectory and the social rationality trajectory (see Table 1). These distinct perspectives generate regulatory debates over proper procedures for risk assessment (that is, type of risk targeted by regulators, the principle of substantial equivalence, appropriate regulatory hurdles and the precautionary principle), risk management (that is, risk tolerance, role of non-scientific information as well as regulatory structure, focus and participation) and, finally, risk communication (that is, role for labelling regulations). The WTO regulatory approach embodied in the SPS Agreement supports the scientific rationality trajectory with its focus on technological progress and product-based regulatory oversight. The Biosafety Protocol approach supports the social rationality trajectory with its focus on technological precaution and process-based regulatory oversight.

Table 1   GMO regulatory trajectories

General regulatory issues	The Risk Analysis Framework (RAF)	
	Scientific rationality	*Social rationality*
Belief	Technological progress	Technological precaution
Type of risk	Recognized	Recognized
	Hypothetical	Hypothetical *and* Speculative
Substantial equivalence	Accepts substantial equivalence	Rejects substantial equivalence
Science or other in risk assessment	Safety Health	Safety Health Quality 'Other legitimate factors'
Burden of proof	Traditional: Innocent until proven guilty	Guilty until proven innocent
Risk tolerance	Minimum risk	Zero risk
Science or other in risk management	Safety or hazard-basis: Risk management is for risk reduction and prevention only	Broader socio-economic concerns: Risk management is for social responsiveness
Specific regulatory issues		
Precautionary principle	Scientific interpretation	Social interpretation
Focus	Product-based, novel applications	Process- or technology-based
Structure	Vertical, existing structures	Horizontal, new structures
Participation	• Narrow, technical experts	• Wide, 'social dimensions'
	• Judicial decision-making	• Consensual decision-making
Mandatory labelling	Safety- or hazard-based strategy	Consumers' right to know based

The differences between the scientific rationality perspective and the social rationality perspective begin with a fundamental difference in the belief about the appropriate role of science and technology in society. According to the former, technology yields innovations and enhances efficiency, which produces economic development and growth, and, in turn produces higher incomes. As incomes go up, demand increases for more stringent social regulations in areas such as food safety and environmental protection. The result is a regulatory race to the top made possible by scientific advancements (Backhouse, 1994; Grossman and Helpman, 1991).

Hence, the goal of this perspective is to set regulatory policies that maximize technological progress, subject to achieving certain standards of safety. Moreover, this foundation creates a regulatory trajectory focused on the novelty of the GMO, not how it was produced. In contrast, the social rationality perspective begins with a much different view on the role of technology in society. Rather than being viewed as objective 'drivers of economic development,' science and technology are viewed as normative activities that by nature bring change to what is a delicate social balance of the preferences and concerns of all constituents. Given that change disrupts the balance, the social rationality perspective supports regulatory policies that ensure technological precaution; if science is going to bring change, then it is important to make sure that *all* impacts of this change are dealt with in a socially responsive manner (Giddens, 1994; Beck, 1992). The result is a pursuit of zero risk which, administratively, means a great deal of analysis and assessment must occur before new technologies are approved. That is, risk minimization is not simply the point where the marginal cost of additional risk reduction activities exceeds their marginal benefit. According to this perspective, progress in science and technology cannot be left to the competitive economic forces of the market. This focus on technological precaution creates a regulatory trajectory focused on the technology or the process of modern biotechnology rather than on the novelty of the GMO.

To elaborate on these different perspectives, consider the role of the precautionary principle within each perspective. While both interpret the precautionary principle as essentially meaning that in the face of uncertainty, when scientific evidence is insufficient, regulators must employ precaution, the two approaches operationalize the principle in fundamentally different ways. The scientific rationality perspective operationalizes the precautionary principle as a risk assessment tool. It is the risk assessors – with their scientific credentials – who can pull the precautionary trigger. When evaluating a new technology there is, of course, an absence of data on the risks and, hence, risks are calculated according to cause–consequence models built from the accumulated peer-reviewed scientific literature. There are two scenarios when the precautionary principle can be invoked. First, suppose there is an absence of scientific literature. Risk assessors would be unable to build their causal–consequence models of risk likelihood and therefore the precautionary principle would be invoked and the technology would not be allowed to proceed through the regulatory review process. Second, when sufficient scientific literature exists and a causal–consequence model can be built precaution is often exercised by specifying risk-averse assumptions or parameters within the likelihood functions, essentially overestimating risk. Therefore, as a scientifically rational risk assessment tool, the precautionary principle is grounded in sound science where the precautionary trigger can only be pulled by risk assessors who hold a required amount of scientific credibility thus producing a *rules-based* approach. The social rationality perspective operationalizes the precautionary principle as both a risk assessment tool *and also* as a risk management tool. According to this perspective, the precautionary principle can be used by risk assessors as above, and in addition, it can also be legitimately employed by risk managers to ensure precaution in the face of non-scientific perceptions and concerns. In other words, using the precautionary principle as a risk management tool potentially increases the social responsiveness of regulations but this also increases the *discretionary* nature of regulations.

Practically speaking, the regulatory approaches to biotechnology products found in the United States and in the co-complainants are consistent with the product-based, novelty-focused scientific rationality perspective. In contrast, the regulatory approaches to agricultural biotechnology products found within the EU are consistent with the process-based, technology-focused social rationality perspective. Transatlantic regulatory regionalism has been created as GMOs approved under the scientific rationality perspective in the US and Canada are delayed or denied access to the EU because it operationalizes the precautionary principle in a manner consistent with the social rationality trajectory. The challenge for the WTO is to identify which interpretation of the principle is consistent with trade rules.

Regulatory decision-making approaches

The focus of the WTO's regulatory decision-making approach is to outline the conditions under which a member can legitimately violate the principle of non-discrimination (PND). That is, the following decision filter is employed (see Table 2). The first question is to ask whether the PPM-based ban is for safety reasons or for non-safety reasons.

If the ban is based on safety-related PPM – the manner in which a good is processed or produced affects the safety of the good – then the international trade regime would deal with the ban in the following manner (the left-hand column of Table 2). A scientific justification must exist to prove that the ban is legitimately safety-related. What constitutes a scientific justification? The WTO itself does not decide scientific legitimacy. As outlined above, it defers to three international scientific organizations whose mandates are to develop both international standards and international standards-setting guidelines in their relevant areas: Codex Alimentarius (food safety), Office of Epizootics (animal safety), and the International Plant Protection Convention (plant safety). If, according to the standards or standards-setting guidelines used by any of the three relevant organizations, a legitimate scientific justification exists for the ban, then the member is permitted under the Agreement on the Application of Sanitary and Phytosanitary Measures (SPS Agreement) to impose the ban in violation of all the concepts of non-discrimination. For instance, 'like' products from one jurisdiction may be banned because of particular safety risks only relevant to that jurisdiction and, in this case, the member does

Table 2   WTO regulatory decision-making approach

WTO approach		
1. Is it a safety-related or non-safety-related ban?		
(i) Safety-related risk assessment	**(ii) Non-safety-related risk management**	
	A. Nature of PPM?	
	(a) Product-related PPM (novelty)	**(b) Non-product-related PPM**
SPS Agreement	**TBT Agreement**	**Out-of-scope**
Requires a scientific justification	Permissible use of TBT measures to ban trade where measures may violate the 'like' products concept of non-discrimination	No permissible use of TBT measures to ban trade (for example, shrimp-turtle, tuna-dolphin)
Permissible use of SPS measures to ban trade where measures may violate all three concepts of the principle of non-discrimination provided a scientific justification exists	However, measures must adhere to the national treatment and most-favoured nation concepts of non-discrimination	Measures must adhere entirely to non-discrimination
(for example, beef hormones case)		

not have to offer either national treatment or most-favoured nation status to that jurisdiction.

If the trade ban is focused on non-safety-related PPM – that is, there is no scientific justification to support a safety-related ban – then a different question is posed to determine legitimacy. What is the nature of the PPM under consideration; are they *product-related* or *non-product-related*? This categorization of PPM was the result of a compromise struck during the Tokyo Round of multilateral trade negotiations under the GATT (Grimwade, 1996). Most countries recognized that in some cases, a product's non-safety-related PPMs were relevant for issues of market access, however, there was considerable disagreement on how to relax the 'like' products concept of non-discrimination in order to deal with this. Countries such as Canada and the United States wanted all technical barriers to trade under the discipline of the international trading system in order to prevent countries from imposing trade bans through mechanisms such as packaging and labelling rules. In essence, this meant that countries could legitimately use non-safety-related PPM measures to ban trade subject to certain constraints. On the other hand, less-developed countries such as India did not want any linking of trade rules and PPM, which meant that there would be no legitimate violations of the 'like' products concept. The compromise was to split PPM into two categories, product-related PPM and non-product-related PPM.

In the event that the PPMs are found to be product-related, the relevant WTO agreement is the Agreement on Technical Barriers to Trade (TBT Agreement) where there are legitimate violations of the 'like' products concept (the middle column of Table 1). For example, consider two meat products where one has been produced under conventional practices while the other has been produced under a quality assurance programme. As a result of the quality assurance programme, the latter product's PPM impacts the quality of the final product and would be considered a product-related PPM. Yet, while in this case there is a legitimate violation of the like products concept, the regulations must still adhere to both the national treatment and the most-favoured nation concepts.

In the event that the PPMs are found to be non-product-related, then they are 'out-of-scope' which means that there are no legitimate circumstances where they may be used to justify a barrier to trade because they cannot legitimately violate any of the three concepts of the principle of non-discrimination. For instance, consider two cotton shirts where the cotton for the first has been grown in an intensive agricultural system while the cotton for the second has been grown in an organic system. Regardless of the intensive or organic PPM employed, there is no impact upon the final product (the cotton shirt) and so the PPM would be considered non-product-related and a trade ban based on the agricultural system used in the production of the cotton would be in contravention of international trade rules.

While the WTO approach is to specify the conditions under which a member can legitimately violate the principle of non-discrimination, the Biosafety Protocol approach to regulatory decision-making is to specify – under the principle of advance informed agreement – the arguments a signatory country can employ when evaluating a biotechnology-based product for domestic market access (see Table 3).

The approach outlined in Table 3 is entirely consistent with the socially rational perspective of the RAF. Of course, there is scope for a scientific risk assessment (left-hand column of Table 3), but this scope is broadened from a classic approach to include a life-cycle analysis of the technology. Such an analysis aims to assess risks across the extraction, production, consumption and disposal stages of a product's life rather than just focusing on the risks associated with the consumption stage. This is a very sensible approach if one considers biotechnology-based products to be equivalent to hazardous or toxic waste. Here it would be entirely reasonable to allow a potential Party of Import to assess the product in terms of its scientific risk.

Recall that under the socially rational regulatory trajectory there is a focus on technological precaution in order to ensure that new technologies are not too disruptive. Such an approach is operationalized under the Biosafety Protocol in the following manner. First, the risk managers can investigate

Table 3   Biosafety Protocol approach to regulatory barriers

Biosafety Protocol approach		
Three types of justification under the principle of advance informed agreement		
Risk assessment: Scientific risk assessment	**Risk management: Risk acceptability**	**Risk management: Socio-economic considerations**
Focus is on process – LMOs Article 3(a to g)	Regulatory trigger: *Are risks domestically acceptable and manageable?*	Regulatory trigger: *What are the likely socio-economic impacts arising from the*
Science-based Article 15(1)	Annex III(e)	*transboundary movement of LMOs?* Article 26(1)
Can use a life-cycle analysis Article 16(4)		

the domestic acceptability of the risk (middle column of Table 3). This steps away from the classic risk analysis approach where risks are managed in a relative fashion (for example, the risk of a GMO relative to the risk of being struck by lightening) and towards an absolute risk management approach. Second, even if the product is deemed safe and the risk is deemed acceptable, another risk management filter may be applied: the socio-economic considerations from market access (right-hand column of Table 3).[12] Accordingly, the product may be investigated in terms of the economic disruption that it may create in the country of import through, for instance, competition with domestic suppliers. Rather than letting the market decide which products will persist, the principle of advance informed agreement potentially allows for an active state role in managing the risk of economic competition. Admittedly, this may be an extreme interpretation. The point here is simply to acknowledge the potential actions of Parties of Import given the provisions outlined in the Protocol. As an example, consider the supply management of agricultural sectors such as dairy. These sectors are often managed for domestic political economy reasons such as to sustain incomes in politically sensitive regions. Past arguments in support of such programmes have ranged from the preservation of rural communities to arguments that supply management achieves an added environmental dividend in the form of the preservation and conservation of biodiversity. It is not hard to imagine such arguments under Article 26(1) of the Protocol.

Again, the regulatory decision-making approach embodied in the Biosafety Protocol is entirely reasonable if one views biotechnology-based products as equivalent to hazardous or toxic waste. It would be necessary to be precautious, to perform a risk assessment on a life-cycle basis, to investigate risk acceptability and socio-economic disruption prior to any imports.

Implications and conclusions

From the analysis above, it is clear that while there appears to be some concert in the regulatory approaches of the World Trade Organization and the Biosafety Protocol, these are dominated by more fundamental areas of conflict. Since 1947 the international trading system has endeavoured to make trade a commercial function while the Biosafety Protocol endeavours to make trade a state–state function. Based upon the principle of non-discrimination, the WTO agreements seek to outline legitimate violations of this principle while the Biosafety Protocol, based on the principle of advance informed agreement, treats biotechnology-based products as equivalent to hazardous or toxic waste. With respect to the RAF, the WTO employs a scientifically rational approach while the Biosafety Protocol employs a socially rational approach. In terms of regulatory decision-making, the WTO employs a step-wise approach to determine the legitimacy of a market access barrier beginning with identifying if the ban is safety-related or not, then the nature of the ban with respect to the product's PPM. In contrast, the Biosafety Protocol – consistent with the view that biotechnology-based products should be considered as risky as toxic or hazardous waste – dramatically broadens the scope of legitimate market access barriers to include risk acceptability and socio-economic considerations.

But does this mean that the two international regulatory frameworks will clash? After all, there are many MEAs with trade impacts that peacefully co-exist with the international trading system. Consider this question in the context of the general insights of *international policy coordination* between trade agreements and MEAs; that they are likely to be in concert when (1) the MEAs have a narrow scope and (2) there exists a transatlantic consensus on the systemic regulatory principles required to deal with the particular issue. Inversely, they are likely to be in conflict when the MEAs have a broad scope and when there exists transatlantic regulatory regionalism; precisely the characteristics present in the WTO–Cartagena Protocol relationship.

Given the differences outlined above, it is clear that the factors conducive to a harmonious trade–environment relationship are not present with respect to the two multilateral paradigms for regulating biotechnology products. Moreover, there appears to be very little likelihood of convergence because it would require either side to abandon their fundamental regulatory approaches. Given the significant commercial trajectory in North America it is unlikely that the regulatory structure will revert to a precautionary, process-based approach that would treat biotechnology-based products as hazardous or toxic wastes. Inversely, given the politicization of the GMO issue, it is unlikely that the EU will undertake a dramatic withdrawal from the precautionary process-based system in favour of widespread market access based on the principle of non-discrimination.

Which international regulatory framework will prevail? This is a very difficult question to answer. On the one hand, it is entirely reasonable to argue that the current EU support for a Biosafety Protocol-type regulatory approach is the result of a commercial lag (it is much easier to adopt a precautionary approach when the GM crops are foreign and not domestic products) but when domestic products are ready a more technologically progressive approach will emerge in the EU which will then influence the Biosafety Protocol. On the other hand, however, it is also entirely reasonable to argue that the North American/WTO approach lacks sufficient social responsiveness and therefore will collapse in the face of consumer concerns about environmental sustainability and corporate control over the food supply.

In the meantime, fragmented international markets remain with some jurisdictions supporting WTO-style market access rules while others support Biosafety Protocol-style market access rules. This fragmentation has the possibility of creating a protracted diplomacy conflict whereby MEAs are negotiated not as a complementary agreement but rather as a countervailing force to trade liberalization agreements.

Notes

1. This dispute is: DS 292 European Communities – Measures Affecting the Approval and Marketing of Biotech Products (http://www.wto.org/english/tratop_e/dispu_e/cases_e/ds292_e.htm).
2. Historically, trade policy has been a subset of foreign policy far removed from domestic concerns and focused squarely on removing border measures such as tariffs and quotas through the rules of international diplomacy (Johnson, 2000; Stairs, 2000; Milner, 1998). Given the general success of trade liberalization – border measures on manufactured goods have fallen steadily – the attention of the international trading regime has increasingly turned to new trade policy challenges including regulatory regionalism and the appropriate relationship between trade and the environment.
3. See Kerr and Hobbs (2002) for a discussion of the beef hormone case.
4. This criticism is contestable. The relationship between international trade agreements and measures to protect the environment has received enough attention that in Articles 31 to 33 of the Doha Agenda's Ministerial Declaration, there are calls for greater clarification on this issue. Despite this attention, however, the actual record has not been that antagonistic at all. In 2001, the WTO's Committee on Trade and the Environment recognized 238 MEAs, 32 of which were deemed to contain trade-distorting provisions (WTO CTE 2001). Three particularly trade-distorting MEAs include: the Convention on the International Trade in Endangered Species 1973 (CITES); the Montreal Protocol on Substances that Deplete the Ozone Layer 1987 (Montreal Protocol); and the Basel Convention on the Transboundary Movement of Hazardous Wastes and Their Disposal 1989 (Basel Convention). Despite their trade-distorting provisions, to date, no MEAs have been directly challenged under the auspices of the WTO. Instead, they have peacefully co-existed.

5. Separate customs territories such as Taiwan and Hong Kong are also allowed to be members.

6. Scientific justifications are determined to be sound not by the WTO but by one of three international scientific agencies that the WTO defers to (1) Codex Alimentarius Commission (food safety and human health), (2) International Office of Epizootics (animal safety and health), and (3) International Plant Protection Convention (plant safety and health).

7. The term used to describe the products of modern biotechnology in the BSP is 'living modified organisms'.

8. Cartagena Protocol on Biosafety to the Convention on Biological Diversity: Text and Annexes 2000). Montreal: Secretariat to the Convention on Biological Diversity (http://www.biodiv.org). Hereinafter referred to as the Biosafety Protocol (2000). Article 4: Scope.

9. See Isaac, Phillipson and Kerr (2002) for a discussion of the tuna-dolphin and shrimp-turtle disputes.

10. The RAF was first codified in 1983 by the US National Academy of Sciences. Science (which meant natural or hard science) was deemed to be a superior baseline for policy-making for two reasons. First, it was argued that natural science strove to disentangle normative dimensions from positive dimensions during the enquiry process in a way just not possible with social sciences. As a result, natural science could produce facts about the actual safety of a product that were not embedded with risk perceptions. Second, it was argued that disagreement in scientific results sets in motion an accepted methodology for debate and reconciliation of results. For instance, if two scientists assessing the actual risk of a product arrived a much different conclusions, then a comparison of the scientific protocols, controls, materials and procedures used is launched. The RAF has three components. The first, risk assessment, is designed to provide (to the extent possible) an objective and neutral product risk profile identifying the actual risk (not the perceived risk). The second component, risk management, is designed to make a regulatory decision based upon the product risk profile established by the risk assessors. Finally, risk communication is designed to ensure transparency; a two-way flow of information between both the risk assessors and the risk managers but also between the RAF and affected stakeholders.

11. GMO regulations based on the RAF can be found in many countries such as the United States, the European Union (as well as in the member states), Canada, Australia and Japan The RAF is also supported by international organizations such as the Organization for Economic Cooperation and Development (OECD), the World Trade Organization (WTO) and several United Nations agencies including the World Health Organization (WHO), the Food and Agriculture Organization (FAO) and the Codex Alimentarius Commission.

12. The allowance for socio-economic considerations further muddies the waters in this case because they cannot be disentangled from traditional protectionist economic interests whose influence on government's decisions trade agreements attempt to specifically limit through their rule-making. Hence, there is a widely held perspective in North America that the EUs approach to the regulation of GMOs simply represents an alternative way for the EU to respond to traditional protectionist influences in their agricultural sector.

12
The Cartagena Protocol on Biosafety and the Development of International Environmental Law

Ruth Mackenzie

Introduction

The period since the mid-1980s has seen a flurry of environmental policy- and law-making at the international level. The 1992 UN Conference on Environment and Development provided the impetus for the adoption of global conventions addressing climate change and biodiversity, and sparked new international negotiations on a range of environmental issues, including desertification, straddling fish stocks and hazardous chemicals. The same period has seen an intensification of debates around the evolution and status of certain principles of international environmental law, particularly the precautionary principle, as well as of the debate over the proper relationship between international trade rules and environmental protection (see Sands, 2003).

This chapter examines aspects of the Cartagena Protocol on Biosafety against the broader context of international environmental law, with a view to exploring the ways in which the Protocol contributes or may contribute to the development of international environmental law in general.[1] Given space limitations, the chapter focuses on a limited set of cross-cutting issues that have broader relevance in international environmental law and that have been, and continue to be, particularly contentious in the context of the Protocol itself. This chapter can offer only a brief survey of these issues, and in some respects it would be premature to seek to do more at this stage. The Protocol has only been in force for a little over 2 years, and the vast majority of Parties are still in the midst of developing national implementing laws and regulations. Moreover the Protocol itself is still in many respects a work in progress: several of its more detailed rules and procedures are still under discussion and remain contentious.[2] With that limitation in mind, this chapter briefly explores the approach and effect of the Protocol in relation to the following issues: the precautionary principle and international trade rules, liability and redress, compliance, and financial assistance and capacity-building.[3]

The Cartagena Protocol on Biosafety

The Biosafety Protocol was negotiated under the auspices of the Convention on Biological Diversity (CBD). It was adopted in January 2000 and entered into force in September 2003. By December 2005, 129 states and the European Communities had become Parties to it.

The structure and approach of the Cartagena Protocol on Biosafety are similar to that of numerous other multilateral environmental agreements (MEAs) adopted in recent years.[4] Among other things, the Protocol sets out certain general principles; it establishes a set of substantive rules, in this case governing transboundary movement of genetically modified organisms (GMOs);[5] in respect of certain issues, it provides for the consideration and development of more detailed rules in the future; it establishes a set of institutional arrangements and procedures to oversee and administer the implementation and evolution of the agreement; and it provides for financial assistance to developing country Parties for implementation. The Protocol is thus situated firmly within a broader corpus of international environmental law. In its subject-matter and approach, it straddles two otherwise rather distinct clusters of MEAs: biodiversity-related treaties, on the one hand, and treaties dealing with the transboundary movement of hazardous substances and products, or the other. The Protocol embodies some of the most challenging issues faced by modern international environmental law. For example, its subject-matter requires balancing of environmental and health concerns with concerns about trade and development; it was negotiated and must be implemented in a context of continuing scientific uncertainty and lack of scientific consensus about the potential risks of GMOs and the manageability of those risks; it raises issues related to who should properly bear the risk of any damage caused by GMOs; and it raises enormous questions about enforcement capacity and the 'governability' of the certain activities. In a number of respects then, the Protocol offers, or it seems likely to demand, novel approaches to challenges faced in other MEAs.

The focus of the Protocol is on the transboundary movement of GMOs. The central procedural mechanism set out in the Protocol to regulate the transboundary movement of GMOs is advance informed agreement (AIA). The AIA procedure essentially requires that before the first transboundary movement of a GMO, the Party of import is notified of the proposed transboundary movement and is given an opportunity to decide, within 270 days, whether or not the import should be allowed and upon what conditions. This decision must be based upon a risk assessment, carried out in a scientifically sound manner, taking into account recognised risk assessment techniques. Article 15 of the Protocol sets out the risk assessment requirements in more detail, and Annex III contains guidance on the objective of risk assessment, the general principles of risk assessment, and the methodology to be applied. The Protocol recognises that risk assessment must be environment-specific, that is, it must consider the risks associated with the release and use of the GMO in the environmental conditions into which it is to be introduced. Where there is a lack of scientific certainty about the extent of the potential adverse effects of a GMO, a Party may take precautionary action to avoid or minimise the potential adverse effects. The Party of import may also take into account certain socio-economic considerations, pursuant to Article 26 of the Protocol, in reaching a decision on the proposed import. However, any such consideration must also be consistent with that Party's other international obligations. In addition, the Biosafety Protocol contains certain obligations regarding public awareness and participation as regards the decision-making process, although the obligation to involve the public in decision-making on GMOs is qualified by a reference to national laws and regulations.

The AIA procedure only applies to the first transboundary movement of a particular GMO into a country for *intentional introduction into the environment* (for example, seeds for open field trials or for commercial growing). Separate, and less onerous, provisions in Article 11 apply to the import of GMOs intended for direct use as food or feed or for processing (essentially agricultural commodities and GMOs used in industrial processes). This procedure, which basically comprises a multilateral information exchange mechanism, centres on the Biosafety Clearing-House (BCH), which was established under Article 20 of the Biosafety Protocol. Parties that authorise domestic use of a GMO that may be used directly as food or feed, or for processing inform other Parties through the BCH; and Parties that require advance notification and approval before the import of such a GMO into their territory alert other Parties and exporters to this fact through the BCH. In effect, the provisions of Article 11 allow Parties to subject GMOs intended for direct use as food or feed, or for processing to similar import requirements to those under the AIA procedure.

The Protocol's AIA procedure resembles the prior informed consent procedures established for hazardous wastes and certain hazardous chemicals and pesticides in other MEAs.[6] However, the approach taken in the Protocol differs in two significant respects. First, the prior informed consent regimes in respect of transboundary movement of hazardous wastes and chemicals apply to categories of substances and materials which it is internationally agreed, through the relevant MEA, are hazardous in nature. By contrast, the Protocol sets out a procedure to be applied on a GMO-by-GMO basis, in the context of specific proposed uses and receiving environments (Stoll, 2000). There still remains some significant disagreement at the international level as to the nature and extent of any special risks that GMOs in general, or any specific GMO, might pose to the environment and human health. This means, of course, that there remains plenty of scope for dispute as to the outcome of the AIA process in the case of any particular proposed import. Secondly, there is a significant degree of flexibility in how the Protocol's procedures, including AIA, are to be applied by Parties. The Protocol confers some discretion upon Parties as to what GMOs are subject to the AIA procedure, and as to whether to follow the AIA procedure itself or some similar domestic regulatory procedure. This flexibility is constrained by the requirement that any such alternative measures must be 'consistent with this Protocol'.

The import and export provisions of the Biosafety Protocol are backed up by requirements setting out what information must be provided in documentation accompanying transboundary movements of GMOs (Article 18). This information is intended to provide a means to identify and track transboundary movements of GMOs; provide information to the Party of import at the border; and offer a contact point for further information about the consignment in question. The specific requirements vary according to the intended use of the GMOs in question. The identification requirements for shipments of GM agricultural commodities remain the most contentious issues (see Falkner and Gupta, 2004).

The Protocol does not prohibit trade in GMOs between Parties and non-Parties, but it requires that such transboundary movements be carried out in a manner 'consistent with the objective' of the Protocol (Article 24). This was the subject of significant debate during the Protocol negotiations since one of the major exporters of GMOs, the United States, is presently not entitled to become a party to the Biosafety Protocol as it has not yet ratified the CBD (Article 32, CBD).

During the negotiation of the Protocol, developing countries pressed for the inclusion of provisions on liability and redress for any damage caused by GMOs. In the event, only an enabling clause was incorporated into the Protocol requiring Parties to adopt a process with respect to the appropriate elaboration of rules and procedures on liability and redress, with a view to completing the process within 4 years (Article 27).

The Biosafety Protocol does not contain specific provisions relating to the settlement of disputes arising under it. Instead, it relies on the relevant provisions of its parent convention, the CBD, which provides for optional judicial or arbitral settlement of disputes or compulsory (at the request of one party), but non-binding, conciliation (Article 27, CBD). In this respect, the Protocol, in common with other MEAs, is significantly weaker than the World Trade Organization agreements. In common with numerous other recent MEAs, the Protocol also provides for the establishment of a non-compliance procedure (Article 34).

The Biosafety Protocol, the precautionary principle and international trade rules

It was clear from the earliest stages of the negotiation of the Biosafety Protocol that its implementation would impact on international trade in

GMOs. A number of trade measures are either required by or authorised under the Protocol. For example, the Protocol establishes an AIA procedure, requiring prior notification and assessment of first import of a GMO, and allowing Parties to restrict or prohibit imports of specific GMOs, or subject them to conditions. It allows countries to take such decisions in reliance on a precautionary approach in circumstances where there is scientific uncertainty due to lack of scientific information and knowledge regarding the extent of potential adverse effects of the GMO. Furthermore, it contains specific requirements relating to the identification (documentation and labelling) of transboundary shipments of GMOs.

States have taken different approaches to the status of the precautionary principle in international law, and no clear and uniform understanding has yet emerged (Sands, 2003: 272). Various formulations of the principle have been included in numerous MEAs and in non-binding declarations in recent years. The principle has also been invoked by some Parties in disputes before certain international courts and tribunals, but it has yet to receive unequivocal recognition as a principle of customary international law by such a body.[7] The inclusion of a reference to precaution in the preamble and objective of the Protocol was relatively uncontroversial, although some states insisted that the reference be to the precautionary *approach* rather than the precautionary *principle*, and that the reference be made specifically to the formulation contained in Principle 15 of the Rio Declaration.[8] The inclusion of some formulation of the precautionary principle in the operative provisions of the Protocol that deal with import decisions (see now Article 10(6) and Article 11(8)), and that could potentially result in trade restrictions, was far more contentious, and represents a significant attempt to put the principle into operation and an advance for the role of the precautionary principle in MEAs.[9] The Protocol now *authorises* Parties expressly to take precautionary decisions in the face of scientific uncertainty regarding the extent of potential adverse impacts of a GMO. When the relevant operative provisions are read together with the objective of the Protocol, it is arguable that the Protocol goes further and *requires* Parties to take a precautionary approach to import decisions in the face of such scientific uncertainty. Nonetheless, the language of the precautionary provisions in Article 10(6) and 11(8) of the Protocol is sufficiently imprecise to give rise to disagreement as to their proper application.

The potential impact of the Protocol on international trade in biotechnology products, together with the explicit incorporation of the precautionary approach in its operative provisions, rendered the question of the Protocol's relationship with international trade rules of enormous potential significance. However, despite protracted negotiations, the issue was not fully resolved within the Protocol (see generally, Koester, 2005). Commentators have taken different views as to the effect of the three preambular paragraphs in the Protocol that address this issue.[10] Some emphasise the relatively clear language of the second of the three preambular paragraphs as evidence of clear intent of the Parties that the Protocol should not be taken as changing rights and obligations of Parties under any existing international agreements, including relevant WTO agreements (Safrin, 2002a). Others suggest that a plain reading of the preambular language fails to resolve the question of the intended relationship between the Protocol and the WTO one way or the other, in that it reflects, and appears to have been intended to reflect, the opposed positions taken throughout the Protocol negotiations (French, 2001: 138; Stoll, 2000).

While the issue of the relationship between MEAs and WTO rules has been the subject of extensive academic enquiry and policy debate, and now formally comprises part of the Doha negotiation mandate,[11] it has not to date arisen squarely in a dispute before the WTO dispute settlement mechanism. However, the Protocol is potentially of relevance in the current WTO dispute between the United States, Argentina and Canada, on the one side, and the European Communities, on the other. None of the complainant WTO members are Parties to the Biosafety Protocol[12] and, to that extent, it is questionable whether the Protocol can be invoked in relation to the objections raised by the complainants relating to the application of the European Union's regulatory system for GMOs. However, the Protocol has been ratified by the European Community and 129 states, and its very adoption would appear to evidence at least a degree of consensus in the international community that GMOs are the appropriate subject of specific regulatory approaches, that they should be subject to careful risk assessment, and that it is appropriate to apply a precautionary approach to their import in situations of scientific uncertainty. The way in which the panel, and potentially the Appellate Body, approach (or avoid) the question of the relevance of the Biosafety Protocol in the dispute seems certain to have significant implications for the broader debate about the MEA–WTO relationship, even if it is not likely definitively to resolve it. It is notable that the Doha mandate in this regard carefully preserves the position of non-Parties to MEAs, but in other cases concerning trade-related environmental measures, the WTO Appellate Body has emphasised the importance of seeking multilateral solutions to transboundary environmental concerns.[13]

The issue of the relationship between the Protocol and the WTO is generally discussed in rather negative terms that focus on the existence and resolution of conflict between the various agreements. There is, however, also the possibility that the Protocol might prove to demonstrate ways in which trade and environment agreements can be mutually supportive. The Protocol, establishes basic standards for import procedures for GMOs, particularly those intended for intentional introduction into the environment. In setting out the AIA procedure, including risk assessment requirements and guidance, it should serve to harmonise national approaches to decision-making on GMO imports. In this respect, and if one accepts the basic premise that there is nothing in the Protocol that *prima facie* requires a Party to act inconsistently with any of its relevant obligations as a WTO member (Charnovitz, 2000; Mackenzie, 2002; cf Stoll, 2000), the Protocol could serve as a useful, more detailed supplement to existing WTO disciplines, such as those contained in the Agreement on the Application of Sanitary and Phytosanitary Measures (Stoll, 2000: 114). As a broader consideration, the intensive implementation and capacity-building effort focused on the Protocol in developing country Parties could have the side effect of strengthening implementation of other sanitary and phytosanitary measures, for example, in relation to alien invasive species and food safety.

Liability and redress

Negotiations over the inclusion of rules on liability and redress in the Biosafety Protocol were fraught with tension between developing and developed countries (see Cook, 2002). The end result was an enabling provision contained in Article 27 of the Protocol that requires Parties to adopt a process with regard to the appropriate elaboration of international rules and procedures on liability and redress with a view to completing the process within 4 years. A Working Group on Liability and Redress has been established (Decision BS-I/8). Its work is at an early stage and there remains disagreement as to what the appropriate outcome of this process should be, and, in particular, whether a binding liability regime should be adopted under the Protocol in respect of liability for damage arising from the transboundary movement of GMOs.

Numerous other instruments have already been developed in international environmental law addressing liability for environmental damage. These have traditionally tended to be sectoral, and stand-alone arrangements, addressing for example, damage from oil pollution and nuclear accidents. Such instruments tend to focus on civil liability approaches – that is, they allocate or 'channel' liability to identifiable private economic actors, principally those in operational control of the activity that causes damage. Supplementary mechanisms may allocate some liability in addition to other entities, for example, in relation to oil pollution, to cargo owners. States have been reluctant to establish liability regimes, which provide for state liability for transboundary environmental damage caused by private activities under their jurisdiction or control.

There has been a growing tendency towards including liability in the agenda of broader MEA negotiations (Brunnée, 2004). For example, in 1999 a Protocol on Liability was adopted under the Basel Convention on the Transboundary Movement of Hazardous Wastes, and in 2003 a Liability Protocol was adopted in the UNECE region in respect of the effects of industrial accidents on transboundary waters. There has also been an attempt on a regional basis to develop general rules on environmental liability in the

Council of Europe Lugano Convention on Civil Liability for Damage Resulting from Activities Dangerous to the Environment, adopted in 1993. The International Law Commission (ILC) has grappled with the question of liability for environmental damage for a number of years. Having finalised Draft Articles on Prevention of Transboundary Environmental Harm in 2001, the ILC turned its attention again to the potential elaboration of a set of generally applicable articles on liability for environmental harm, again opting for an approach focused principally on civil liability.[14]

While the negotiations on liability and redress under the Protocol can benefit from the extensive work already conducted at the international level, they appear to be taking place amid an atmosphere of some fatigue and growing scepticism about the value of international liability regimes for environmental damage (see, for example, Brunnée, 2004; Daniel, 2003). Daniel (2003) points to the length of time taken to negotiate such arrangements and to prospects of entry into force. The Lugano Convention, mentioned above, has not entered into force, having failed to attract a single ratification since 1993. The Protocol on Liability to the Basel Convention has attracted only 7 accessions and 13 signatures, from among the Convention's 166 Parties since it was adopted in 1999 after several years of negotiation.[15] Of the liability regimes addressing environmental damage that have been negotiated at the international level, only those in respect of oil pollution and nuclear damage are in force (Brunnée, 2004). Daniel suggests that efforts would be better directed at other mechanisms, such as: strengthening national laws designed to prevent and punish environmental harm; capacity-building; or the use of private international law mechanisms or international funds (Daniel, 2003: 238–40). Brunnée similarly questions whether liability regimes are an appropriate tool for environmental protection, noting that only the oil pollution agreements have so far been successfully relied upon by pollution victims (Brunnée, 2004). She further suggests that the experience of the oil pollution regimes is not necessarily a good indicator that such regimes are appropriate within the context of other MEAs, particularly for those, such as biosafety, in which any damage may involve diffuse sources, cumulative effects and long time lags (Brunnée, 2004).

Despite such reservations, it seems unlikely that the calls by many developing country Parties to the Protocol for a liability regime will diminish in the near future. There is a general sense among such Parties that such a regime would more fairly allocate the risks associated with the transfer and use of GMOs, and would provide a degree of comfort to those countries and communities that have concerns about the potential impacts of GMOs in the environment. However, any liability regime developed under the Protocol seems likely to have to find novel solutions to certain difficult issues that have arisen in relation to environmental liability generally. These include the problem of the definition of environmental damage,[16] particularly damage to biological diversity, and the appropriate way to channel liability in the context of the Protocol's AIA-focused approach, which assumes that states importing LMOs will have had an opportunity to assess fully possible risks and to select appropriate risk management measures. In this respect, the traditional tendency to focus on 'operator' liability (that is, allocating liability primarily to the person in 'operational control' of the activity or material) may not satisfy developing country concerns. Some proposals made in the discussions on this issue to date have tended to focus rather on the possibility of liability of the Party of export or on the exporter of the GMO, a solution unlikely to be acceptable to developed countries. The AIA context also raises questions about the scope of any liability regime: should the regime only cover damage caused during the transboundary movement of the GMO or should it also extend to subsequent use of that GMO in the importing Party?

While the allocation of liability and the scope of any regime are matters that will have to be tailored very specifically to the context of the Protocol, any agreement reached on the definition of damage to be covered might well have implications for other areas of international environmental law. Most liability regimes adopted to date have struggled with the definition of environmental damage, generally limiting it to costs of measures to prevent such damage or to costs of reinstatement (see generally, Bowman and Boyle, 2002). A few have gone further and sought to cover environmental damage *per se*, but significant conceptual problems have arisen with regard to valuation of such damage. The definition of damage to biological diversity seems likely to provide even more challenging, particularly in the context of GMOs.

Compliance

In common with many recent MEAs, the Biosafety Protocol provides for the establishment of a compliance procedure. Sands (2003: 203) has identified the emergence of non-compliance mechanisms under various MEAs, with a function between conciliation and traditional dispute settlement, as one of the most significant recent developments in international environmental law. Compliance procedures in MEAs generally exist alongside, and operate without prejudice to, traditional mechanisms for the settlement of disputes concerning the interpretation and application of the agreement. The dispute settlement provision of the CBD applies to the Biosafety Protocol and, in common with many other MEAs, that provision is fairly weak insofar as there is only recourse to binding third-party dispute settlement (that is, arbitration or reference to the International Court of Justice) if both Parties to the dispute so agree.[17] Despite some concerns in the negotiation of the Protocol regarding its possible relationship with international trade rules, there was no serious consideration of establishing any binding and mandatory dispute resolution procedures within the Protocol itself (Mackenzie and Sands, 2002: 464).

Compliance procedures under MEAs are generally primarily facilitative in nature. They are designed to address cases of non-compliance and assist and advise Parties to resolve problems meeting their obligations under the agreement in question, including through the provision of financial and technical assistance. Article 34 of the Biosafety Protocol, which mandates the establishment of such a procedure, refers to 'cooperative procedures and institutional mechanisms'. However, compliance procedures with a harder edge have been adopted, notably under the Kyoto Protocol to the UN Framework Convention on Climate Change. The Kyoto compliance procedure comprises both a facilitative and an enforcement branch, with the latter able to adopt specific responses to certain types of non-compliance with Kyoto commitments (see Wang and Wiser, 2002). Despite, or perhaps because, of the vexed question of the Protocol's relationship with international trade rules, there was no such innovation in the compliance procedure adopted under the Cartagena Protocol at the first meeting of the Parties in 2004 (Mackenzie, 2004: 272–3; see Decision BS-I/7). Instead the Protocol's compliance procedure resembles more those adopted under other MEAs such as the Montreal Protocol on Substances that Deplete the Ozone Layer and the Basel Convention. The issue of whether or not some type of sanction might be imposed in response to certain cases of non-compliance was extremely contentious, and was deferred until the third meeting of the Parties in 2006, but no further decision was taken on this issue at that meeting.[18]

It remains to be seen how the compliance procedure will operate in practice and what degree of 'activism' the members of the compliance committee will adopt when or if they are faced with specific compliance questions that require interpretation of provisions of the Protocol. There are a number of aspects of the Protocol that remain open to interpretation and in respect of which no specific oversight mechanism has been established. For example, the Protocol requires that Parties conduct transboundary movements of GMOs to and from non-Parties 'consistent with the objective of this Protocol' (Article 24), and that if a Party elects to utilise a procedure other than AIA for imports of GMOs then that procedure be 'consistent with this Protocol' (Article 9). No oversight procedures are established in respect of these provisions and, as yet, there is no authoritative interpretation from the Protocol's governing body as to what types of standards or procedures would satisfy these requirements.

The Protocol's compliance procedure is also more restrictive when viewed against the innovation in the procedure adopted under the UNECE Aarhus Convention on Access to Information, Public Participation in Decision-Making and Access to Justice in Environmental Matters. That procedure allows members of the public to initiate complaints regarding alleged non-compliance by a Party to the Convention. As regards access to the compliance procedure, while Article 23 of the Biosafety Protocol apparently evidences a degree of commitment to public participation at the national level, this is not backed up with any rights to initiate or otherwise formally participate in the compliance procedure. The possibility is left open that the public, or non-governmental organisations, may provide information to the compliance committee, insofar as the list of those who can make information available is non-exhaustive (Decision BS-I/7, para. V.2).

Financial resources and capacity-building

Article 22 of the Protocol on capacity-building is not dissimilar to other provisions on this issue found in other recent MEAs. It has increasingly been recognised, particularly since UNCED in 1992, that developing country Parties require specific financial assistance in order to implement their commitments under MEAs, and that efforts should be made to build technical, human and institutional capacity to address environmental problems. This is especially the case where the perceived environmental threat posed is potentially of a global nature, is novel, demands specific expertise and may be costly to address. Article 22 sets outs a general obligation to cooperate in capacity-building for biosafety, including biotechnology to the extent it is required for biosafety. Financial resources are addressed in Article 28, which links the Protocol to the CBD's financial mechanism, operated by the Global Environment Facility (GEF). In effect, GEF funds are available for implementation of the Protocol by developing country Parties, within the provisions of the CBD and the governing instruments of the GEF.

To what extent does the Protocol offer any innovation in respect of capacity-building and financial assistance? While commitments in respect of capacity-building and finance in MEAs, and in the Protocol itself, have tended to be rather general, in practice under the Protocol there has been at least an attempt to take a more strategic approach in terms of their implementation. To a large degree this is dictated by the nature of the Protocol, which is predicated upon a system of working national regulatory frameworks for biosafety. The functioning of the AIA procedure requires that countries have established a competent national authority, that they can assess information regarding a GMO for potential import, that they can reach import decisions and implement any risk management measures deemed necessary, and that they can monitor actual effects of the GMO after its release into the environment. Of course, this is far from being the case in most of the countries that have ratified the Protocol. Indeed, concerns about the lack of such frameworks were at the heart of developing country demands for a Biosafety Protocol to be elaborated. Given the scale of growth in commercial cultivation in GMOs in recent years (see James, 2005a), growth in the transboundary movement of GMOs, interest in GM crop development as a potential contribution to addressing food security problems in developing countries, and concerns over the potential environmental, health and socio-economic impacts of GMOs, there was a strong perception among many developing countries in the Protocol negotiations that rapid progress towards the development of natio nal regulatory frameworks was urgently needed.

To address this need, the GEF adopted an 'Initial Strategy for assisting countries to prepare for the entry into force of the Protocol on Biosafety'. Part of this strategy was a GEF-funded project implemented by UNEP on the Development of National Biosafety Frameworks.[19] The $38.4 million project, which began in 2001, was intended to assist up to 100 countries to develop their national biosafety frameworks (NBFs). Well over 100 countries are now participating in this project and more than 50 have completed their NBF. Development of the NBF is however, only a first step and does not mean that a country is in a position to begin to deal with notifications of proposed imports in accordance with the Protocol's AIA provisions. Many countries are in the process of applying to the GEF for additional funds for implementation projects to put the NBF into practice. Alongside the UNEP-GEF and related projects, there are a host of other multilateral, regional and bilateral initiative underway which aim, in some way, to build capacity in biosafety, or which focus on modern biotechnology (for example, GM crop research and development) but comprise a biosafety component.

The adoption of the Protocol has thus seen perhaps the most extensive efforts in any MEA to rapidly bring countries to a situation where they are capable of implementing their obligations under the agreement. These efforts have not been without difficulty: issues of coordination and of targeting appropriate assistance have arisen. There are at least three major challenges faced by capacity-building efforts in this field: first, it is widely recognised that to build necessary internal capacity in many countries to undertake risk assessment of GMOs is a task that will take many years, extending beyond the life of most existing capacity programmes; second, in many countries the capacity-building efforts began in a virtual policy vacuum: thus policy development, public awareness, elaboration of national regulations and building scientific and technical capacity have been attempted simultaneously; and third, some capacity-building initiatives, and efforts by developing countries to elaborate national rules and procedures in respect of GMO imports, have been affected by broader international developments, including an awareness of the ongoing dispute between the US and the EU in the WTO.

Nonetheless, the approach to capacity-building taken in respect of the Biosafety Protocol seems to represent an approach previously untried in international environmental law. Experiences and lessons learned, both good and bad, seem likely to have an impact on how such initiatives are financed and designed in the future in respect of other environmental issues.

Conclusions

At the time the Cartagena Protocol on Biosafety was negotiated, the issues it addressed in terms of risks associated with the transfer and use of GMOs remained relatively novel. To some degree the novelty of the issues being addressed appears to have prompted innovation in the design and implementation of the Protocol. For example, the Protocol adopts an explicitly precautionary approach; and it has given rise to rapid and extensive capacity-building initiatives. On the other hand, relative novelty, and the Protocol's potential trade impacts particularly over the longer term, may have impeded innovation, for example, in relation to compliance. The differing perspectives of countries as to risks associated with modern biotechnology, and the differing levels of existing national regulation, have given rise to a multilateral agreement that leaves potentially enormous scope for national discretion in the extent and manner of regulation of the transboundary movement of GMOs. This flexibility, coupled with the absence of strong dispute settlement or other oversight procedures, may prove a weakness in the overall effectiveness of the Protocol as a global instrument to protect biological diversity from any adverse effects associated with GMOs.

It seems premature to attempt any definitive assessment of the contribution of the Protocol to international environmental law in the fields of liability for environmental damage and compliance. As noted above, discussions on liability remain at a relatively early stage, and negotiations on the content of any liability regime under the Protocol have not really begun in earnest. The elaboration of a meaningful liability regime under the Protocol would seem to require the Parties to address some novel issues, such as the definition of damage to biological diversity, that could have broader implications for other environmental liability regimes. As regards compliance, the design of the procedure and mechanisms under the Protocol has not involved a significant degree of innovation as compared to similar regimes under MEAs. Any attempt to draw conclusions as to the contribution of the compliance mechanism to international environmental law more generally should await some practice of the consideration by the Compliance Committee of specific cases of alleged non-compliance.

At present the primary contribution of the Protocol to international environmental law in general appears to be in respect of the precautionary principle. The integration of precaution explicitly into a decision-making procedure on imports of GMOs and GM commodities represents a major step in international environmental law, and a significant challenge in the understanding of the relationship between trade and environmental agreements. It may have implications for future agreements addressing environmental, and perhaps also health, risks, particularly as national experience grows of implementation of these provisions of the Protocol. This experience will also inform evolution of the wider understanding of the meaning and status of the principle in international environmental law.

The adoption and entry into force of the Protocol have brought into sharper focus questions relating to the proper relationship between trade and environmental agreements, and have increased the likelihood that at some stage the WTO dispute settlement procedure will have to address this issue more squarely. It has further heightened questions and concerns about the proper role of WTO dispute settlement procedure in assessing the WTO consistency of national trade measures affecting imports of GMOs, in a context in which there remains a lack of international consensus about appropriate regulatory approaches and outcomes. In this regard, the fact that some of the most significant exporters of GMOs and GM commodities, in particular the US, remain outside the Protocol represents a particular challenge for reconciling the trade and environment relationship.

Notes

1. An initial attempt to examine certain aspects of this issue was made in Mackenzie and Sands (2002)
2. For a discussion of some of the key outstanding issues, see Falkner and Gupta (2004) and Mackenzie (2004).
3. There are a number of other issues that might also form part of this enquiry. These include, for example, the nature and content of prior informed consent procedures in international environmental law, and the role of science in the evolution and implementation of multilateral environmental agreements. For reason of limitations of space, these issues are outside the scope of the present chapter. The issue of the role of science in the Protocol has been considered in Redgwell (2005). Redgwell also considers the approach taken in the Protocol to public participation in decision-making on GMOs. More detailed analyses of some of the issues addressed in this chapter are referred to in the relevant sections and references that follow.
4. There was significant reference throughout the negotiations to precedents and examples from other MEAs, particularly those in the field of hazardous substances: the 1989 Basel Convention on the Control of Transboundary Movements of Hazardous Wastes and Their Disposal and the 1998 Rotterdam Convention on the Prior Informed Consent Procedure for Certain Hazardous Chemicals and Pesticides in International Trade. The institutional and financial provisions draw upon the 1998 Kyoto Protocol to the UN Framework Convention on Climate Change; and the provision in Article 35 on the review of effectiveness of the Protocol draws upon a similar provision in the 1987 Montreal Protocol on Substances that Deplete the Ozone Layer. While negotiators drew on existing examples, they also had in mind the possible impact that approaches and language adopted in the Biosafety Protocol might have on the future negotiation of other MEAs. In particular, a number of the Biosafety Protocol negotiators were at the same time involved in the negotiation of the 2001 Stockholm Convention on Persistent Organic Pollutants.
5. While the Protocol refers to 'living modified organisms', in this chapter the more commonly used terms 'genetically modified organisms' or 'GMOs' are utilised throughout.
6. See note 4 above.
7. In the *Beef Hormones* case in the WTO, the Appellate Body, considering the potential relevance of the precautionary principle in a case involving human health concerns, observed that it was not clear that the precautionary principle had crystallised into a principle of customary international environmental law. *EC Measures Concerning Meat and Meat Products (Hormones)*, Report of the Appellate Body, 16 January 1998, Doc. WT/DS26/AB/R and Doc. WT/DS48/AB/R.

In the *Southern Bluefin Tuna* cases, the International Tribunal for the Law of the Sea, making a provisional measures order, did not explicitly endorse the precautionary principle, but nonetheless referred to the need for the Parties to 'act with prudence and caution' to ensure that effective conservation measures were taken to prevent serious harm to southern bluefin tuna stocks, and referred in this context to the problem of scientific uncertainty regarding appropriate conservation measures. *Southern Bluefin Tuna Cases*, (New Zealand v Japan; Australia v Japan), Order on Provisional Measures, 27 August 1999), paras. 77–81.

8. Principle 15 provides:

In order to protect the environment, the precautionary approach shall be widely applied by states according to their capabilities. Where there are threats of serious or irreversible damage, lack of full scientific certainty shall not be used as a reason for postponing cost-effective measures to prevent environmental degradation.

9. Some other MEAs do contain precautionary approaches in their operative provisions, for example, the 1995 Agreement on Straddling and Highly Migratory Fish Stocks and certain marine pollution agreements (Mackenzie and Sands, 2002: 462).

10. The solution drew upon the example of the 1998 Rotterdam Convention. The preambular paragraphs read as follows:

Recognizing that trade and environment agreements should be mutually supportive with a view to achieving sustainable development, *Emphasizing* that this Protocol shall not be interpreted as implying a change in the rights and obligations of a Party under any existing international agreements, *Understanding* that the above recital is not intended to subordinate this Protocol to other international agreements.

11. Paragraph 31 of the Doha Declaration provides (in part)

With a view to enhancing the mutual supportiveness of trade and environment, we agree to negotiations, without prejudging their outcome, on: the relationship between existing WTO rules and specific trade obligations set out in multilateral environmental agreements (MEAs). The negotiations shall be limited in scope to the applicability of such existing WTO rules as among Parties to the MEA in question. The negotiations shall not prejudice the WTO rights of any Member that is not a party to the MEA in question . . .

12. Argentina and Canada have signed the Protocol.

13. *United States – Import Prohibition of Certain Shrimp and Shrimp Products*, Report of the Appellate Body, 12 October 1998, WTO Doc. WT/DS58/AB/R, para. 166.

14. The text of the *Draft articles on the allocation of loss in the case of transboundary harm arising out of hazardous activities* was adopted on first reading by the International Law Commission in 2004 and transmitted to governments for comments and observations by 1 January 2006. *Report of the International Law Commission*, 56th Session, UN Doc. A/59/10. See Boyle (2005). In 2005, an Annex on Liability arising from Environmental Emergencies was adopted under the Protocol on Environmental Protection to the Antartic Treaty.

15. The Basel Liability Protocol has a relatively low threshold for entry into force, requiring 20 ratifications or accessions.

16. The ways in which existing liability regimes in international environmental law address the difficult question of the definition of environmental damage has been analysed in de La Fayette (2002).

17. The CBD does provide for a non-binding conciliation procedure at the request of one party if the dispute is not settled by other means.
18. This matter was again deferred. It is also notable that the third meeting of the Parties also failed to agree on a rule for decision-making by the Compliance Committee by majority or qualified majority vote. Thus, unlike other compliance mechanisms, for the time being at least, the Protocol's Committee can only act by consensus. Both issues remain pending for further consideration by the meeting of the Parties in the future.
19. Further information on this project is available at http://www.unep.ch/bioafety.

References

AATF (2002) *Rationale and Design of the AATF*. African Agricultural Technology Foundation website, June. Available at http://www.aatf-africa.org/rationale.php.

ACDI/VOCA (2003) *Genetically Modified Food: Implications for US Food Aid Programs*, 2nd ed., April. Available at www.acdivoca.org/acdivoca/acdiweb2.nsf/news/gmfoodsarticle.

Action Aid (2003) *GM Crops: Going Against the Grain* (London: Action Aid).

Aerni, P. (2001) 'Assessing Stakeholder Attitudes to Agricultural Biotechnology in Developing Countries', *The Biotechnology and Development Monitor* 47: 2–7.

Aerni, P. (2002a) 'Stakeholder Attitudes towards the Risks and Benefits of Agricultural Biotechnology in Developing Countries: A Comparison between Mexico and the Philippines', *Risk Analysis* 22(6): 1123–37.

Aerni, P. (2002b) *Public Attitudes towards Agricultural Biotechnology in South Africa*. STI/CID Research Report (Cambridge: Center for International Development, Harvard University). Available at http://www.iaw.agrl.ethz.ch/~aernip/PDF/SAreport.pdf.

Aerni, P. (2003) *The Private Management of Public Trust: The Changing Nature of Political Protest*. Working document for the European Consortium of Political Research (Zurich). Available at http://www.essex.ac.uk/ECPR/events/generalconference/marburg/papers/4/5/Aerni.pdf.

Aerni, P. (2004) '10 Years of Cassava Research at ETH Zurich – A Critical Assessment', in ZIL Annual Report 2003: 20–7. Available at http://www.zil.ethz.ch/docs/annual_reports/AnnualReportZIL2003.pdf.

Afonso, M. (2002) 'The Relationship with Other International Agreements: An EU Perspective', in C. Bail, R. Falkner and H. Marquard (eds) *The Cartagena Protocol on Biosafety: Reconciling Trade in Biotechnology with Environment and Development?* (London: Earthscan), pp. 423–37.

African Centre for Biosafety, Earthlife Africa, Environmental Rights Action-Friends of the Earth Nigeria, Grain and SafeAge (2004) *GE Food Aid: Africa Denied Choice Once Again?* 4 May. Available at http://www.biosafetyafrica.net/_DOCS/Africa_GM_food_aid.pdf.

African Centre for Biosafety (2005) *African Biosafety Laws and Comments*. Available at http://www.biosafetyafrica.net/biosafety_laws_and_comments.htm.

Agence Europe (2002) 'Parliament Strengthens Labelling Rules on GM Food but Avoids Upsetting Balance', 3 July.

Agricultural Biotechnology Support Project II (ABSPII) (2004) *Scope and Activities*, ABSPII website. Available at http://www.absp2.cornell.edu/whatisabsp2/.

AIBA (2000) *Biotechnology Parks*, AIBA Report (Delhi: AIBA).

Alden, E. (2000) 'Greens and Free-Traders Join to Cheer GM Crop Deal', *The Financial Times*, 31 January.

Alden, E. (2003) 'US Beats Egypt with Trade Stick', *The Financial Times*, 30 June.

Anderson, K. and S. Yao (2003) 'China, GMOs and World Trade in Agricultural and Textile Products', *Pacific Economic Review* 8(2): 157–69.

Anderson, R., S. E. Levy and B. M. Morisson (1991) *Rice Science and Development Politics: Research Strategies and IRRI's Technologies Confront Asian Diversity (1950–1980)* (Oxford: Clarendon Press).

Andrée, P. (2005) 'The Cartagena Protocol on Biosafety and Shifts in the Discourse of Precaution', *Global Environmental Politics* 5(4): 25–46.

Angelo, I., F. Masiga and L. Musiita (2003) 'Africa's Dilemma in Genetically Modified Food War', *The Monitor* (Kampala), 29 May.

Arts, B. (1998) *The Political Influence of Global NGOs, Case studies on the Climate and Biodiversity Conventions* (Utrecht: International Books).

Arts, B. and S. Mack (2003) 'Environmental NGOs and the Biosafety Protocol. A Case Study on Political Influence', *European Environment* 13(1): 19–33.

Arts, B. and P. Verschuren (1999) 'Assessing Political Influence in Complex Decision-Making. An Instrument Based on Triangulation', *International Political Science Review* 20(4): 411-24.

AstraZeneca (1999) *Annual Report* (AstraZeneca).

Bail, C. (2000) 'The Convention on Biological Diversity in an European Perspective', *Bulletin de L'Institut Royal des Sciences Naturelles de Belgique: Biologie* 70(Suppl.): 21–3.

Bail, C., J. P. Decaestecker and M. Jørgensen (2002) 'European Union', In C. Bail, R. Falkner and H. Marquard (eds) *The Cartagena Protocol on Biosafety: Reconciling Trade in Biotechnology with Environment and Development?* (London: Earthscan), pp. 166–85.

Bail, C., R. Falkner and H. Marquard (eds) (2002) *The Cartagena Protocol on Biosafety: Reconciling Trade in Biotechnology with Environment and Development?* (London: Earthscan).

Baron, D. (2002) *Business and Its Environment*. 4th ed. (Upper Saddle River, NJ: Prentice Hall).

Barrett, C. and D. Maxwell (2005) *Food Aid After Fifty Years: Recasting Its Role* (London: Routledge).

Barrett, K. and E. Abergel (2000) 'Genetically Engineers Crops: Breeding Familiarity: Environmental Risk Assessment for Genetically Engineered Crops in Canada', *Science and Policy* 27(1): 2–12.

Bauer, M. (ed) (1995) *Resistance to New Technology: Nuclear Power, Information Technology and Biotechnology*. Cambridge: Cambridge University Press.

Bauer, M. W., G. Gaskell and J. Durant (eds) (2002) *Biotechnology: The Making of a Global Controversy* (Cambridge: Cambridge University Press).

Baumgartner, F. R. and B. D. Jones (1993) *Agendas and Instability in American Politics* (Chicago, IL: University of Chicago Press).

Baumüller, H. (2003) *Domestic Import Regulations for Genetically Modified Organisms and Their Compatibility with WTO Rules*. Trade Knowledge Network (International Institute for Sustainable Development and International Centre for Trade and Sustainable Development). Available at http://www.tradeknowledgenetwork.org/pdf/tkn_domestic_regs.pdf.

Baumüller, H. (2004) 'Domestic Import Regulations for Genetically Modified Organisms and Their Compatibility with WTO Rules. *Asian Biotechnology and Development Review* 6(3): 33–42.

Beck, U. (1992) *Risk Society: Towards a New Modernity* (London: Sage Press).

Beck, U. (1999) *World Risk Society* (Cambridge: Polity Press).

Bender, K. L. and R. E. Westgren (2001) 'Social Construction of the Market(s) for Genetically Modified and Nonmodified Crops' *American Behavioral Scientist* 44(8): 1350–70.

Benedick, R. E. (1991) *Ozone Diplomacy: New Directions in Safeguarding the Planet* (Cambridge: Harvard University Press).

Bennett, J. (2003) 'Food Aid Logistics and the Southern Africa Emergency', *Forced Migration Review* 18(5).

Bernasconi-Osterwalder, N. (2001) 'The Cartagena Protocol on Biosafety: A Multilateral Approach to Regulate GMOs', In E. B. Weiss and J. H. Jackson (eds) *Reconciling Environment and Trade* (Ardsley, NY: Transnational Publishers, Inc.).

Bernauer, T. (2003) *Genes, Trade and Regulation: The Seeds of Conflict in Food Biotechnology*, Princeton: Princeton University Press.

Bernauer, T. and E. Meins. (2003) 'Technological Revolution Meets Policy and the Market: Explaining Cross-National Differences in Agricultural Biotechnology Regulation', *European Journal of Political Research* 42(5): 643–83.

Betsill, M. (2001) 'NGO Influence in International Environmental Negotiations: Desertification and Climate Change', *Global Environmental Politics* 1(4): 65–84.

Beyers, J. (2002) 'Gaining and Seeking Access: The European Adaptation of Domestic Interest Associations', *European Journal of Political Research* 41(5): 585–613.

Bijman, J. (2001) 'AgrEvo: From Crop Protection to Crop Production', *AgBioForum* 4(1): 20–5.

Biotechnology Industry Organization (BIO) (1999) *Biosafety Protocol – An Overview.* Available at http://www.bio.org/food&ag/bspoverview.html.

Bisang, R. (2003) 'Diffusion Process in Networks: The Case of Transgenic Soybean in Argentina', Paper read at Conferência Internacional sobre Sistemas de Inovação e Estratégias de Desenvolvimento para o Terceiro Milênio, Brazil (November).

Black, J. (1998) 'Regulation as Facilitation: Negotiating the Genetic Revolution', *Modern Law Review* 61(5): 621–60.

Black, J. (2002) *Critical Reflections on Regulation.* CARR Discussion Paper Series. (London).

Blackhouse, R.E. (1994) *Economists and the Economy: The Evolution of Economic Ideas.* 2nd ed. (New Brunswick, NJ: Transaction Publishers).

Blassnig, R. (2000) 'Countries Agree on Biosafety Protocol Regulating Transboundary Movement of GMOs', *International Environment Reporter* 23(3) (2 February).

Bob, C. (2002) 'Merchants of Morality', *Foreign Policy* (129) (March/April): 36–45.

Bonny, S. (2003) 'Why are Most Europeans Opposed to GMOs? Factors Explaining Rejection in France and Europe', *Electronic Journal of Biotechnology.* Available at http://www.ejbiotechnology.info/content/vol6/issue1/full/4/bip/.

Borlaug, N. (2003) 'Science vs. Hysteria', *The Wall Street Journal*, 22 January.

Bowman, M. and A. E. Boyle (eds) (2002) *Environmental Damage in International and Comparative Law: Problem of Definition and Valuation* (Oxford: Oxford University Press).

Boyle, A. E. (2005) 'Globalising Environmental Liability: The Interplay of National and International Law', *Journal of Environmental Law* 17: 3–26.

Brac de la Perriere, R. A. and F. Seuret (2000) *Brave New Seeds: The Threat of GM Crops to Farmers* (London: Zed Books).

Brack, D., R. Falkner and J. Goll (2003) *The Next Trade War? GM Products, the Cartagena Protocol and the WTO.* Briefing Paper No. 8 (London: Chatham House).

Breslin, S (2003) 'Reforming China's Embedded Socialist Compromise: China and the WTO', *Global Change, Peace and Security* 15(3): 213–29.

Bruninga, S. (2000) 'Industry Applauds Biosafety Protocol That Allows for More Information Sharing', *International Environment Reporter*, 23(4) (16 February).

Brunnée, J. (2004) 'Of Sense And Sensibility: Reflections On International Liability Regimes As Tools For Environmental Protection', *International and Comparative Law Quarterly* 53: 351–68.

Burchell, J. and S. Lightfoot (2004) 'Leading the Way? The European Union at the WSSD', *European Environment* 14: 331–41.

Byrne, D. (2000a) *Biotechnology: Building Consumer Acceptance*. Speech given at the European Business Summit. Available at http://europa.eu.int/comm/dgs/health_consumer/library/speeches/speech49_en.html.

Byrne, D. (2000b) *Food Safety A Top Priority in the EU*. Speech given at the Agriculture Council, Biarritz. Available at http://europa.eu.int/comm/dgs/health_consumer/library/speeches/speech54_en.html.

Caduff, L. (2005) *Vorsorge oder Risiko? Umwelt- und verbraucherschutzpolitische Regulierung im europäisch-amerikanischen Vergleich: Eine politökonomische Analyse des Hormonstreits und der Elektronikschrott-Problematik* (Center for International and Comparative Studies, ETH Zürich: PhD thesis).

Carroll, R. (2002) 'Zambia Slams Door Shut on GM Food Relief', *The Guardian* (London), 30 October.

Cashore, B. (2002) 'Legitimacy and the Privatization of Environmental Governance', *Governance* 8(4): 503–29.

Cathie, J. (1997) *European Food Aid Policy* (Aldershot: Ashgate).

CEC (2000) *Communication from the Commission on the Precautionary Principle*. Commission of the European Communities. Available at http://europa.eu.int/comm/dgs/health_consumer/library/pub/pub07_en.pdf.

CEC (2002) *Towards a Strategic Vision of Biotechnology and Life Sciences*. Commission of the European Communities. Available at http://europa.eu.int/comm/biotechnology.

CEC (2003) *Towards a Strategic Vision of Biotechnology and Life Sciences: Progress Report and Future Orientations*, 5 March. Commission of the European Communities.

Center for Food Safety (2000) *Take Genetically Engineered Bovine Growth Hormone Off the Market!*, Available at http://www.foodsafetynow.org/send.asp?cam_id=57.

Centro Latinoamericano de Ecología Social (CLAES) (2003) Observatorio Ambiental gropecuario Mercosur 48(4). Available at http://ambiental.net/boletines/index.html.

Centro Latinoamericano de Ecología Social (CLAES) (2004) Observatorio Ambiental gropecuario Mercosur 49(2). Available at http://ambiental.net/boletines/index.html.

Chan, G. (2004) 'China's Compliance in Global Environmental Affairs', *Asia Pacific Viewpoint* 45(1): 69–86.

Charlton, M. (1992) *The Making of Canadian Food Aid Policy* (Montreal: McGill-Queens).

Charnovitz, S. (2000) 'The Supervision of Health and Biosafety Regulation by World Trade Rules', *Tulane Environmental Law Journal* 13(2): 271–303.

Charnovitz, S. (2002) 'Improving the Agreement on Sanitary and Phytosanitary Standards', in G. L. Sampson and W. B. Chambers (eds) *Trade, Environment, and the Millennium*, 2nd ed. (Tokyo: United Nations University Press).

Chase, B. (2000) 'Biotech Crops Stunted by Perception; Consumers Manufacturers Are Slow to Accept Genetically Altered Foods', *Milwaukee Journal Sentinel*, 4 September.

Chataway, J. (2001) 'Novartis: New Agribusiness Strategy' *AgBioForum* 4(1): 14–9.

Chatterjee, P. and M. Finger (1994) *The Earth Brokers. Power, Politics and World Development*, (London: Routledge).

China Business Review (2002) 'China's WTO implementation efforts' *China Business Review* 29(4) July–August.

China Daily (2004) 'Debate Over GMOs Simmers in China', *China Daily*, 6 April.

China Daily (2005) 'China Approves New GMO Cotton to Raise Output', *China Daily*, 19 September.

Christensen, C. (2000) 'The New Policy Environment for Food Aid: The Challenge of Sub-Saharan Africa', *Food Policy* 25(3): 255–68.

Christian Aid (2000) *Selling Suicide: Farming, False Powers and Genetic Engineering in Developing Countries* (London: Christian Aid).

Chudnovsky, D. (2004) *Trade, Environment and Development: The Recent Argentine Experience*. (Brasilia: Working Group on Development and Environment in the Americas).

Clapp, J. (2003) 'Transnational Corporate Interests and Global Environmental Governance: Negotiating Rules for Agricultural Biotechnology and Chemicals', *Environmental Politics* 12(4): 1–23.

Clapp, J. (2004) 'WTO Agricultural Trade Battles and Food Aid', *Third World Quarterly* 25(8): 1439–52.

Clapp, J. (2005) 'Transnational Corporations and Global Environmental Governance', in P. Dauvergne (ed) *Handbook of Global Environmental Politics* (Cheltenham: Edward Elgar).

Clay, E. and O. Stokke (eds) (1991) *Food Aid Reconsidered: Assessing the Impact on Third World Countries* (London: Frank Cass).

Clay, E. and O. Stokke (eds) (2000) *Food Aid and Human Security* (London: Frank Cass).

Clearing House Mechanism/CHM (2000) *Cartagena Protocol on Biological Safety*. Available at http://www.biodiv.org/biosafe/Protocol/Protocol.html.

Clegg, S. (1989) *Frameworks of Power* (London: SAGE).

CMHT/Cohen, Milestein, Hausfeld, & Toll, PLLC (2001) *Case Watch: Biologically Engineered Seeds*. Available at http://www.cmht.com/casewatch/cases/cwstar-link.htm.

Codex Alimentarius Commission (2003) *Report of the Fourth Session of the Ad Hoc Intergovernmental Task Force of Foods Derived from Biotechnology*, ALINORM 02/34A (Rome: FAO and WHO).

Cohen, J. and R. Paarlberg. (2004) 'Unlocking Crop Biotechnology in Developing Countries – A Report from the Field', *World Development* 32(9): 1563–77.

Colborn, T., D. Dumanoski and J. Myers (1996) *Our Stolen Future* (New York: Dutton).

Coleman, W. D. and M. Gabler (2002) 'Agricultural Biotechnology and Regime Formation: A Constructivist Assessment of the Prospects', *International Studies Quarterly* 46(4): 481–506.

Colitt, R. (2003) 'Washington Takes the Battle over Future of Genetically Modified Crops to Brazil', *The Financial Times*, 20 June.

Cook, K. (2002) 'Liability: No Liability, No Protocol', in C. Bail, R. Falkner and H. Marquard (eds) *The Cartagena Protocol on Biosafety: Reconciling Trade in Biotechnology with Environment and Development?* (London: Earthscan), pp. 371–84.

Cooper, C. M. (2006) '"This is Our Way In": The Civil Society of Environmental NGOs in South-West China', *Government and Opposition* 41(1): 109–36.

Correll, E. (1999) *The Negotiable Desert: Expert Knowledge in the Negotiations of the Convention to Combat Desertification* (Linkoping: PhD thesis).

Council of the European Union (1999) *Draft Council Conclusions on the Biosafety Protocol*. Report No. 13344/99 ENV 409. 6 December.

Cox, R. and H. Jacobson (eds) (1973) *The Anatomy of Influence* (New Haven: Yale University Press).

Crilly, R. (2002) 'Children Go Hungry as GM Food Rejected', *The Herald* (Glasgow), 30 October.

Cutler, C., V. Haufler and T. Porter (1999) *Private Authority and International Affairs* (Albany: SUNY).

Dahl, R. A. (1961) *Who Governs? Democracy and Power in an American City* (New Haven: Yale University Press).

Dale, G. (2001) 'Merging Rivulets of Opposition: Perspectives of the Anti-Capitalist Movement', *Millennium* 30(2): 365–79.

Daniel, A. (2003) 'Civil Liability Regimes as a Complement to Multilateral Environmental Agreements: Sound International Policy or False Comfort?' *Review of European Community and International Environmental Law* 12: 225–41.

Dauenhauer, K. (2003) 'Health: Africans Challenge Bush Claim That GM food is Good for Them', *SUNS: South-North Monitor* (No. 5368), 23 June.

Dawkins, K. (1991) *Sharing Rights and Responsibilities for the Environment: Assessing Potential Roles for Non-Governmental Organizations in International Decision-Making* (Massachusetts Institute of Technology).

de La Fayette, L. (2002) 'The Concept of Environmental Damage in International Liability Regimes', in M. Bowman and A. E. Boyle (eds) *Environmental Damage in International and Comparative Law: Problem of Definition and Valuation* (Oxford: Oxford University Press), pp. 150–89.

Denny, C. and L. Elliott (2003) 'French Plan to Aid Africa Could be Sunk by Bush', *The Guardian*, 23 May.

Depledge, J. (2000) 'Rising From the Ashes: The Cartagena Protocol on Biosafety', *Environmental Politics* 9(2): 156–62.

DeSombre, E. R. (2005) 'Understanding United States Unilateralism: Domestic Sources of U.S. International Environmental Policy', in R. S. Axelrod, D. L. Downie and N. J. Vig (eds) *The Global Environment: Institutions, Law, and Policy* (Washington, D.C.: CQ Press).

Días, A. (2001) *Biotecnología, Argentina y los Argentinos* (Buenos Aires) Available at http://www.argiropolis.com.ar/BioPercepcion.htm.

Dicken, P., P. F. Kelly, K. Olds and H. W.-C. Yeung (2001) 'Chains and Networks, Territories and Scales: Towards a Relational Framework for Analysing the Global Economy', *Global Networks* 1(2): 89–112.

Diven, P. (2001) 'The Domestic Determinants of US Food Aid Policy', *Food Policy* 26(5): 455–74.

Doern, G. B. (2000) *Inside the Canadian Biotechnology Regulatory System: A Closer Exploratory Look*, Canadian Biotechnology Advisory Committee, November.

Douglas, M. (1992) *Risk and Blame: Essays in Cultural Theory* (London: Routledge).

Dratwa, J. (2004) 'Social Learning with the Precautionary Principle at the European Commission and the Codex Alimentarius', in B. Reinalda and B. Verbeek (eds) *Decision Making within International Organizations* (London: Routledge).

Dryzek, J. S., D. Downes, C. Hunold, D. Schlosberg and H.-K. Hernes (2003) *Green States and Social Movements: Environmentalism in the United States, United Kingdom, Germany, and Norway* (Oxford, Oxford University Press).

Dynes, M. (2002) 'Africa Torn between GM Aid and Starvation,' *The Times* (London), 6 August: 12.

Earth Negotiation Bulletin (1996a) 'A Summary Report on the Second Session of the Conference of the Parties to the Convention on Biological Diversity', 6–17 November 1995, 9(39).

Earth Negotiation Bulletin (1996b) 'Summary of the First Meeting of the Open-Ended Ad Hoc Working Group on Biosafety', 22–26 July 1996, 9(48).

Earth Negotiation Bulletin (1997) 'Report of the Third Meeting of the Open-Ended Ad Hoc Working Group on Biosafety', 13–17 October 1996, 9(74).

Earth Negotiation Bulletin (1998) 'Report of the Fourth Session of the Open-Ended Ad Hoc Working Group on Biosafety', 5–13 February 1998, 9(85).

Earth Negotiation Bulletin (2005) 'Summary of the First Meeting of the Ad Hoc Group on Liability and Redress and the Second Meeting of the Parties to the Cartagena Protocol on Biosafety', 25 May–3 June 2005, 9(320).

EC (1997a) 'Regulation 97/258/EC of 27 January 1997 Concerning Novel Foods and Novel Food Ingredients', *Official Journal of the European Communities*, L 43 (14 February): 1–6.

EC (1997b) 'Commission Decision of 23 July 1997 Setting up Scientific Committees in the Field of Consumer Health and Food Safety', *Official Journal of the European Communities*, L 237 (28 July): 18–23.

EC (2001) 'European Parliament and Council Directive 2001/18/EC of 12 March on the Deliberate Release into the Environment of Genetically Modified Organisms and Repealing Council Directive 90/220/EEC', *Official Journal of the European Communities* L 106 (17 April): 1–38.

EC (2002) 'Regulation 178/2002 28 January Laying Down the General Principles and Requirements of Food Law, Establishing the European Food Safety Authority, and Laying Down Procedures in Matters of Food Safety', *Official Journal of the European Communities L 31* (1 February): 1–24.

EC (2003a) 'Regulation 1829/2003 of 22 September 2003 on Genetically Modified Food and Feed', *Official Journal of the European Communities* L 268 (18 October): 1–23.

EC (2003b) 'Regulation 1830/2003 of 22 September 2003 Concerning the Traceability and Labelling of GMOs and Traceability of Food and Feed Produced from GMOs and Amending Directive 2001/18', *Official Journal of the European Communities* L 268 (18 October): 24–8.

Economist Magazine (1996) 'Pop Goes the Treaty', 3 August.

Economy, E. C. (1997) 'Chinese Policy-Making and Global Climate Change: Two-Front Diplomacy and the International Community', in M. A. Schreurs and E. C. Economy (eds) *The Internationalization of Environmental Protection* (Cambridge: Cambridge University Press), pp. 19–41.

Economy, E. C. (2004) *The River Runs Black: The Environmental Challenge to China's Future* (Ithaca: Cornell University Press).

Economy, E. C. and Schreurs, M. A. (1997) 'Domestic and International Linkages in Environmental Politics', in M. A. Schreurs and E. C. Economy (eds) *The Internationalization of Environmental Protection* (Cambridge: Cambridge University Press), pp. 1–18.

EEC (1990) Council Directive 90/220 on the Deliberate Release to the Environment of Genetically Modified Organisms, *Official Journal of the European Communities*, L 117, 8 May: 15–27.

EFSA GMO Panel (2004) *Guidance Document of the Scientific Panel on Genetically Modified Organisms for the Risk Assessment of Genetically Modified Plants and Derived Food and Feed*, European Food Safety Authority, September. Available at http://www.efsa.eu.int.

ENDS Daily (1999) 'Outline for EU Public Health Agency Emerges', *ENDS Environment Daily*, 13 December.

Ernst and Young (2002) *Regions of the Future: Life Sciences*. Available at http://www.ey.nl/download/publicatie/Regions_of_the_Future_Life__Sciences.pdf.

ETC Group (2001) 'Globalization, Inc.', *Communique* (71). Available at http://www.rafi.org/documents/com_globalization.pdf.

ETC Group (2002) 'Sterile Harvest: New Crop of Terminator Patents Threatens Food Sovereignty', News Release, 31 January. Available at http://www.etcgroup.org.

ETC Group (2005a) 'Global Seed Industry Concentration – 2005', in *Communique* (90). Available at http://www.etcgroup.org/documents/Comm90GlobalSeed.pdf.

ETC Group (2005b) 'Suicide Seeds – Bombshell in Bangkok', News Release, 11 February. Available at http://www.etcgroup.org/documents/NRTerminator05 Bangkok.pdf.

Eurobarometer (2003) 'A Report to the EC Directorate General for Research from the project 'Life Sciences and European Society'', by G. Gaskell, N. Allum and S. Stares. Available at http://europa.eu.int/comm/public_opinion/archives/eb/ebs_177_en.pdf.

European Parliament (1997) *Report by the Temporary Committee of Inquiry Into BSE on Alleged Contraventions or Maladministration in the Implementation of Community Law in Relation to BSE, Without Prejudice to the Jurisdiction of the Community and National Courts* HE 220.544/fin.

European Union (1995) EU Council Regulation (ED) No. 1292/96.

European Union (2003) Press Release, IP/03/681, 13 May.

European Union (2004) *Economic Review of EU Mediterranean Partners*. Occasional Paper No. 6, Brussels: DG Economic and Financial Affairs. Available at http://europa.eu.int/comm/economy_finance/publications/occasional_papers/2004/ocp6en.pdf.

Evans, Peter (1995) *Embedded Autonomy: States and Industrial Transformation* (Princeton: Princeton University Press).

Falkner, R. (2000) 'Regulating Biotech Trade: The Cartagena Protocol on Biosafety', *International Affairs* 76(2): 299–313.

Falkner, R. (2001) 'Business Conflict and U.S. International Environmental Policy: Ozone, Climate, and Biodiversity', in P. G. Harris (ed) *The Environment, International Relations, and U.S. Foreign Policy* (Washington, D.C.: Georgetown University Press), pp. 157–77.

Falkner, R. (2002a) 'Negotiating the Biosafety Protocol: The International Process', in C. Bail, R. Falkner and H. Marquard (eds) *The Cartagena Protocol on Biosafety: Reconciling Trade in Biotechnology with Environment and Development?* (London: Earthscan), pp. 3–22.

Falkner, R. (2002b) 'International Trade Conflicts over Agricultural Biotechnology', in A. Russell and J. Vogler (eds.) *The International Politics of Biotechnology: Investigating Global Futures* (Manchester: Manchester University Press), pp. 142–56.

Falkner, R. (2004a) 'Tracing Food: The Politics of Genetically Modified Organisms', in B. Hocking and S. McGuire (eds) *Trade Politics*, 2nd ed. (London: Routledge), pp. 249–60.

Falkner, R. (2004b) 'The First Meeting of the Parties to the Cartagena Protocol on Biosafety', *Environmental Politics* 13(3): 635–41.

Falkner, R. (2005) 'American Hegemony and the Global Environment', *International Studies Review* 7(4): 585–99.

Falkner, R. and A. Gupta (2004) *Implementing the Biosafety Protocol: Key Challenges*, Briefing Paper, Sustainable Development Programme, BP 04/04 (London: Chatham House).

FAO (2004a) *The State of Food and Agriculture 2003–2004. Agricultural Biotechnology, Meeting the Needs of the Poor?* (Rome: Food and Agricultural Organization).

FAO (2004b) *FAO Statistical Database*. Available at http://faostat.fao.org.

Federal Register (2001) 'Premarket Notice Concerning Bioengineered Foods; Proposed Rule', 18 January, 56 4706–38.

Feld, W. and R. Jordan 1983) *International Organizations. A Comparative Approach* (New York: Praeger).

Fewsmith, J. (2001a) 'The Political and Social Implications of China's Accession to the WTO', *The China Quarterly* (167): 573–91.

Fewsmith, J. (2001b) *Elite Politics in Contemporary China*. (Armonk, NY: M.E. Sharpe).

Finger, M. and J. Kilcoyne (1997) 'Why Transnational Corporations are Organizing to "Save the Global Environment"', *The Ecologist* 27(4).

Finnemore, M. and K. Sikkink (1998) 'International Norm Dynamics and Political Change', *International Organization* 52(4): 887–917.

Fold, N. (2000) 'Globalization, State Regulation and Industrial Upgrading of the Oil Seed Industries in Malaysia and Brazil', *Singapore Journal of Tropical Geography* 21(3): 263–78.

Food Aid Convention (1999). *Food Aid Convention* (London: International Grains Council).

Food and Drug Administration (FDA) (2000) *Report on Consumer Focus Groups on Biotechnology*, 20 October. Available at http://www.cfsan.fda.gov/~comm/biorpt.html.

Foudin, A. S. (2000) Presentation at the International Food Policy Research Institute, Washington, D.C., 13 November.

French, D. (2001) 'The International Regulation of Genetically Modified Organisms: Synergies and Tensions in World Trade', *Environmental Liability* 9(3): 127–39.

Frey, B. S. and G. Kirchgässner (2002) *Theorie und Anwendung* (München: Vahlen).

Friedmann, H. (1982) 'The Political Economy of Food: The Rise and Fall of the Postwar International Food Order,' in M. Burawoy and T. Skocpol (eds) *Marxist Inquiries: Studies of Labour, Class and States*, supplement to American Journal of Sociology 88: S248–86.

Frieman, W. (1994) 'International Science and Technology and Chinese Foreign Policy' in T. W. Robinson and D. Shambaugh (eds) *Chinese Foreign Policy: Theory and Practice* (Oxford: Clarendon Press), pp. 158–93.

Friends of the Earth International (FOEI) (2003) *Playing with Hunger: The Reality Behind the Shipment of GMOs as Food Aid* (Amsterdam: FOEI, April). Available at http://www.foei.org/publications/pdfs/playing_with_hunger2.pdf.

Fuchs, D. A. (2004) 'Channels and Dimensions of Business Power in Global Governance', Paper presented at the ISA Annual Conference, Montreal, 17–21 March.

Fukuyama, F. (2004) *State-Building: Governance and World Order in the 21st Century* (Ithaca, NY: Cornell University Press).

Gaisford, J. D. and W. A. Kerr (2001) *Economic Analysis for International Trade Negotiations: The WTO and Agricultural Trade* (Northhampton, MA: Edward Elgar).

Gaisford, J. D. and W. A. Kerr (2003) 'Deadlock in Geneva: The Battle over Export Subsidies in Agriculture', *International Economic Journal* 17(3): 1–17.

Galbraith, J. K. (1984) *The Anatomy of Power* (Boston: Houghton Mifflin).

Garcia-Johnson, R. (2000) *Exporting Environmentalism: U.S. Multinational Chemical Corporations in Brazil and Mexico* (Cambridge, MA: MIT Press).

Garrett, G. and S. McCall (1999) 'The Politics of WTO Dispute Settlement', Paper presented at the annual meeting of the American Political Science Association, Atlanta, GA.

Gaskell, G. (2004) 'Science Policy and Society: the British Debate over GM Agriculture', *Current Opinion in Biotechnology* 15: 241–45.

Gaugitsch, H. (2002) 'Scientific Aspects of the Biosafety Debate', in C. Bail, R. Falkner and H. Marquard (eds) *The Cartagena Protocol on Biosafety: Reconciling Trade in Biotechnology with Environment and Development?* (London: Earthscan), pp. 83–91.

Genetic Resources Action International (GRAIN) (2002) 'Better Dead than GM Fed?' *Seedling* (October). Available at http://www.grain.org/seedling/seed-02-10-6-en.cfm.

Gereffi, G. (1994) 'The Organization of Buyer-Driven Global Commodity Chains: How U.S. Retailers Shape Overseas Production Networks', in G. Gereffi and M. Korzeniewicz (eds) *Commodity Chains and Global Capitalism* (Westport, CT: Greenwood Press).

Giddens, A. (1994) 'Living in a Post-Traditional Society', in U. Beck, A. Giddens and S. Lash (eds) *Reflexive Modernism* (Cambridge: Polity Press).

Gill, S. and D. Law (1989) 'Global Hegemony and the Structural Power of Capital', *International Studies Quarterly* 33: 475–500.

Gilpin, R. (2001) *Global Political Economy: Understanding the International Economic Order* (Princeton: Princeton University Press).

Glickman, D. (1999) *New Crops, New Century, New Challenges*, Presentation to the National Press Club, 13 July, pp. 1–7. Available at http://www.usda.gov/news/releases/1999/07/0285.

Global Industry Coalition (2000) 'Biosafety Protocol: A Major Step Forward to Sustain Development of Biotechnology', Press Release. Canada Newswire, 29 January.

Glover, D. (2003a) 'GMOs and the Politics of International Trade,' *Democratising Biotechnology: Genetically Modified Crops in Developing Countries Briefing Series*, Briefing 5, (Brighton, UK: Institute of Development Studies).

Glover, D. (2003b) *Public Participation in National Biotechnology Policy and Biosafety Regulation*. IDS Working Paper 198, August (Brighton: IDS).

Glover, D., J. Keeley, P. Newell and R. McGee (2003) *Public Participation and the Biosafety Protocol*. Commissioned Study for DFID and UNEP-GEF (Brighton: IDS).

Glover, D. and P. Newell (2004) 'Business and Biotechnology: Regulation of GM Crops and the Politics of Influence', in K. Jansen and S. Vellema (eds) *Agribusiness and Society: Corporate Responses to Environmentalism, Market Opportunities and Public Regulation* (London: Zed Books), pp. 200–31.

Goldsmith, E. and J. Mander (eds) (2001) *The Case Against the Global Economy: And for a Turn Towards Localization* (London: Earthscan).

Gomez Mera, L. (2005) 'Explaining Mercosur's Survival: Strategic Sources of Argentine-Brazilian Convergence', *Journal of Latin American Studies* 37(1): 109–40.

Goverde, H., C. Cerny, M. Haugaard and H. Lentner (eds) (2000) *Power in Contemporary Politics* (London: SAGE).

Grassley, C. (2003) 'Salvation of Starvation? GMO Food Aid to Africa', Remarks of Senator Chuck Grassley to the Congressional Economic Leadership Institute, 5 March. Available at http://www.useu.be/Categories/Biotech/Mar0503Grassley Biotech.html#Grassley.

Greenpeace (2000) 'Time to Take Action at Biosafety Negotiations', 27 January.

Greenpeace UK (2002) 'Statements on the Southern African Food Crisis', November.

Grimwade, N. (1996) *International Trade Policy: A Contemporary Analysis* (Routledge: London).

Grossman, G. M. and E. Helpman (1991) *Innovation and Growth in the Global Economy* (Cambridge: MIT Press).

Guardian (1999) 'GM Investors Told to Sell Their Shares', August 25.

Guivant, J. S. (2002) 'Heterogeneous and Unconventional Coalitions Around Global Food Risks: Integrating Brazil into the Debates', *Journal of Environment Policy and Planning* 4(3): 231–45.

Gunningham, N. (1995) 'Environment, Self-Regulation and the Chemical Industry: Assessing Responsible Care', *Law & Policy* 17: 57–109.

Gupta, A. (2000) 'Governing Biosafety in India: The Relevance of the Cartagena Protocol', Discussion Paper 2000–24. Environment and Natural Resources Program, Kennedy School of Government, Harvard University, Cambridge, MA.

Gupta, J. and M. Grubb (eds) (2000) *Climate Change and European Leadership: A Sustainable Role for Europe?* (Dordrecht: Kluwer).

Haas, P. M. (1992) 'Obtaining International Environmental Protection through Epistemic Consensus', in I. Rowlands and M. Green (eds) *Global Environmental Change and International Relations* (Macmillan: London).

Haas, P. M. (1995) 'Epistemic Communities and the Dynamics of International Environmental Co-operation', in V. Rittberger (ed) *Regime Theory and International Relations* (Oxford: Clarendon Press), pp. 168–201.

Hammond, B. and R. Fuchs (1999) 'Safety Evaluation for New Varieties of Food Crops Developed through Biotechnology' in J. Thomas (ed) *Biotechnology and Safety Assessment.* 2nd ed. (London: Taylor and Francis).

Hardin, R. (2002) *Trust and Trustworthiness* (New York: Russell Sage).

Harlander, S. K. (2002) 'Safety Assessments and Public Concern for Genetically Modified Food Products: The American View', *Toxicologic Pathology* 30(1): 132–34.

Hartley, S. and G. Skogstad (2005). 'Regulating Genetically Modified Crops and Foods in Canada and the United Kingdom', *Canadian Public Administration* 48(3).

Hartnell, G. (1996) 'The Innovation of Agrochemicals: Regulation and Patent Protection', *Research Policy* 25(3).

Helleiner, G. (2001) 'Markets, Politics, and Globalization: Can the Global Economy Be Civilized?' *Global Governance* 7(3): 243–63.

Helm, C. (2000) *Economic Theories of International Environmental Cooperation* (Northhampton, MA: Edward Elgar Press).

Helmuth, L. (2000) 'Both Sides Claim Victory in Trade Pact', *Science*, 287(5454).

Herring, R. J. (2005) 'Miracle Seeds, Suicide Seeds and the Poor: Mobilising around Genetically Modified Organisms in India' in R. Ray and M. F. Katzenstein (eds) *Social Movements in India: Poverty, Power and Politics* (Lanham: Rowman and Littlefield).

Ho, P. (2001) 'Greening Without Conflict? Environmentalism, NGOs and Civil Society in China', *Development and Change* 32(5): 893–921.

Hochstetler, K. (2002) 'After the Boomerang: Environmental Movements and Politics in the La Plata River Basin', *Global Environmental Politics* 2(4): 35–57.

Hochstetler, K. (2003) 'Fading Green: Environmental Politics in the Mercosur Free Trade Agreement and Beyond', *Latin American Politics and Society* 45(4).

Hochstetler, K. (2004) *Civil Society in Lula's Brazil.* Working Paper CBS-57-04 (Oxford: Centre for Brazilian Studies).

Hogenboom, B. (1998) *Mexico and the NAFTA Environment Debate. The Transnational Politics of Economic Integration* (Utrecht: International Books).

Hogue, C. (1998a) 'Debate at Biosafety Protocol Talks to Center on Advance Agreement Regime', *International Environment Reporter* 21(17), 19 August.

Hogue, C. (1998b) 'Key Provisions of Biosafety Protocol Left for Final Negotiations in February 1999', *International Environment Reporter* 21(18), 2 September.

Hooghe, L. and G. Marks (2003) 'Unraveling the Central State, but How? Types of Multi-Level Governance', *American Political Science Review* 97(2): 233–43.

Hopkins, R. (1992) 'Reform in the International Food Aid Regime: The Role of Consensual Knowledge', *International Organization* 46(1): 225–64.

Hopkins, R. (1993) 'The Evolution of Food Aid: Towards a Development-First Regime,' in V. Ruttan (ed) *Why Food Aid?* (Baltimore: Johns Hopkins University Press).

Huang, J., R. Hu, C. Pray and S. Rozelle (2004) 'Plant Biotechnology in China: Public Investments and Impacts on Farmers', in New Directions for a Diverse Planet. Proceeding of the 4th International Crop Science Congress, 26 September–1 October 2004, Brisbane, Australia.

Huang, J. and S. Rozelle (2003) 'The Impact of Trade Liberalization on China's Agriculture and Rural Economy' *SAIS Review* 23(1): 115–31.

Huang, J. and Q. Wang (2002) 'Agricultural Biotechnology Development and Policy in China', *AgBioForum* 5(4): 122–35.

Huang, J. and Q. Wang (2003) *Biotechnology Policy and Regulation in China*. IDS Working Paper 195 (Brighton: Institute of Development Studies).

Huberts, L. (1989) 'The Influence of Social Movements on Government Policy', *International Social Movement Research* 2, pp. 395–426.

Huberts, L. and J. Klennijenhuis (1994) *Methoden van invloedsanalyse (Methodologies for the Analysis of Influence)* (Meppel: Boom).

Hughes, C. (1997) 'Globalization and Nationalism: Squaring the Circle in Chinese IR Theory', *Millennium* 26(1): 103–24.

IDEC (2005) *IDEC Pede a Lula Veto da Nova Lei de Biosegurança*. Available at http://ww.idec.org.br.

Institute for the Study of International Migration (2004) *Genetically Modified Food in the Southern African Food Crisis of 2002–2003* (Georgetown: University School of Foreign Service).

International Chamber of Commerce (1997) *A Precautionary Approach: An ICC Business Perspective*. Available at http://www.iccwbo.org/home/statements_rules/statements/1997/buspes.asp

International Grains Council (IGC) (2004). *IGC Annual Report* (London: IGC).

International Seed Federation (2003) *Position on Genetic Use Restriction Technologies* (Bangalore, June) Available at http://www.worldseed.org/Position_papers/Pos_GURTs.htm

International Wheat Council (1991) *The Food Aid Convention of the International Wheat Agreement* (London: International Wheat Council).

Isaac, G. E. (2002) *Agricultural Biotechnology and Transatlantic Trade: Regulatory Barriers to GM Crops* (Oxford: CABI Publishing).

Isaac, G. E. (2003) 'The WTO and the Cartagena Protocol: International Policy Coordination or Conflict?', *Current Agriculture, Food & Resource Issues* 4: 152–9.

Isaac, G. E. and W. A. Kerr (2003a) 'Genetically Modified Organisms and Trade Rules: Identifying Important Challenges for the WTO', *The World Economy* 26(1): 29–42.

Isaac, G. E. and W. A. Kerr (2003b) 'Genetically Modified Organisms at the World Trade Organization: A Harvest of Trouble', *Journal of World Trade* 37(6): 1083–95.

Isaac, G. E., M. Phillipson and W. A. Kerr (2002) *International Regulation of Products of Biotechnology*. Estey Centre Research Paper Number 2. (Saskatoon, Canada: Estey Centre for Law and Economics in International Trade).

James, C. (2000) *Preview: Global Review of Commercialized Transgenic Crops: 2000*. Report No. 21 (Ithaca, NY: ISAAA).

James, C. (2004) *Global Status of Commercialized Biotech/GM Crops 2004: Executive Summary (Preview)*. ISAAA Briefs No. 32 (Ithaca, NY: ISAAA).

James, C. (2005a) *Executive Summary of Global Status of Commercialized Biotech/GM Crops: 2005*, ISAAA Briefs No. 34 (Ithaca, NY: ISAAA).

James, C. (2005b) *Preview: Global Status of Commercialized Biotech/GM Crops: 2005*, ISAAA Briefs No. 34 (Ithaca, NY: ISAAA).

Jansen, K. and E. Roquas (2005) 'Absentee Expertise: Science Advice for Biotechnology Regulation in Developing Countries' in M. Leach, I. Scoones and B. Wynne (eds) *Science and Citizens: Globalization and the Challenge of Engagement* (London: Zed Books), pp. 142–53.

Jasanoff, S. (1995) 'Product Process or Programme: Three Cultures and the Regulation of Biotechnology' in M. Bauer (ed) *Resistance to New Technology: Nuclear Power,*

Information Technology and Biotechnology (Cambridge: Cambridge University Press).

Jasanoff, S. (2005) *Designs on Nature: Science and Democracy in Europe and the United States* (Princeton: Princeton University Press).

Jepson, W. E. (2002) 'Globalization and Brazilian Biosafety: The Politics of Scale over Biotechnology Governance', *Political Geography* 21(7): 905–25.

Johnson, P. M. (2000) 'Beyond Trade: The Case for a Broadened International Governance Agenda', *Policy Matters* 1(3).

Kaufman, M. (2001) 'EPA Rejects Biotech Corn as Human Food', *Washington Post*, 28 July: A2.

Keeley, J. (2003a) *The Biotechnology Developmental State? Investigating the Chinese Gene Revolution.* IDS Working Paper, Biotechnology Policy Series No. 6 (Brighton: Institute of Development Studies).

Keeley, J. (ed) (2003b) *Democratising Biotechnology.* IDS Briefings (Brighton: Institute of Development Studies).

Keeley, J. (2003c) *Regulating Biotechnology in China: The Politics of Biosafety.* IDS Working Paper 208 (Brighton: Institute of Development Studies).

Kelemen, R. D. (2001) 'The Limits of Judicial Power: Trade-Environmental Disputes in the GATT/WTO and the EU', *Comparative Political Studies* 34(6): 622–50.

Kellerhals, M. D. (2001) *USAID Launches Biotechnology Initiatives with Africa: Programs Foster Improved Regulation, Research, Development.* (Washington, D.C.: Office of International Information Programs, US Department of State), 2 March.

Kent, A. (2002) 'China's International Socialization: The Role of International Organizations', *Global Governance* 8(3): 343–64.

Keohane, R. O. (1984) *After Hegemony: Cooperation and Discord in the World Political Economy.* (Princeton: Princeton University Press).

Keohane, R.O., and H.V. Milner (eds) (1996) *Internationalization and Domestic Politics.* Cambridge: Cambridge University Press.

Kerr, W. A. and J. E. Hobbs (2002) 'The North American-European Union Dispute Over Beef Produced Using Growth Hormones: A Major Test for the New International Trade Regime', *The World Economy* 25(2): 283–96.

Killinger, S. (2000) *International Environmental Externalities and the Double Dividend.* (Northhampton, MA: Edward Elgar Press).

Kingdon, J. (1984) *Agendas, Alternatives, and Public Policies* (New York: Harper Collins Publishers).

Kneen, B. (1999) *Farmageddon* (Gabriola, B.C.: New Society).

Koester, V. (2001) 'Cartagena Protocol: A New Hot Spot in the Trade-Environment Conflict', *Environmental Policy and Law* 31(2).

Koester, V. (2005) 'The Cartagena Protocol on Biosafety: A New Hot Spot in the Trade-Environment conflict?' in R. Melendez-Ortiz and V. Sanchez (eds) *Trading in Genes: Development Perspectives on Biotechnology, Trade and Sustainability* (London: Earthscan), pp. 171–98.

Koller, P. (1991) 'Facetten der Macht', *Analyse und Kritik* 13: 107–33.

Kremer M. and A. P. Zwane (2005) 'Encouraging Private Sector Research for Tropical Agriculture', *World Development* 33(1): 87–105.

Kuyek, D. (2000) 'Lords of Poison: The Pesticide Cartel', *Seedling* (June). Available at http://www.grain.org/publications/jun00/jun003.htm.

Kuyek, D. (2002) 'Past Predicts the Future – GM Crops and African Farmers,' *Seedling* (October). Available at www.grain.org/seedling/seed-02-10-3-en.cfm.

Lapegna, P. and D. Domínguez (2004) 'Transformaciones en la Estructura Social Agraria Argentina. El Problema de la Biotecnología Aplicada al Agro', Paper read at Latin American Studies Association, XXV International Congress, 7–9 October, at Las Vegas.

Lapitz, R., G. Evia and E. Gudynas (2004) *Soja y Carne en el Mercosur: Comercio, Ambiente y Desarrollo Agropecuario* (Montevideo: Coscoroba).

La Viña, A. G. M. (2002) 'A Mandate for a Biosafety Protocol: The Jakarta Negotiations', in C. Bail, R. Falkner and H. Marquard (eds) *The Cartagena Protocol on Biosafety: Reconciling Trade in Biotechnology with Environment and Development?* (London: Earthscan), pp. 34–43.

Lean, G. (2000) 'Rejected GM Food Dumped on the Poor', *The Independent* (London), 18 June.

Levidow, L. (1998) 'Democratizing Technology – or Technologizing Democracy? Regulating Agricultural Biotechnology in Europe', *Technology in Society* 20: 211–26.

Levidow, L. (1999) 'Regulating Bt maize in the United States and Europe: A Scientific-Cultural Comparison', *Environment* (December): 10–23.

Levidow, L. and Bijman, J. (2002) 'Farm Inputs under Pressure from the European Food Industry', *Food Policy* 27(1): 31–45.

Levidow, L. and S. Carr (2000) 'Unsound Science? Transatlantic Regulatory Disputes over GM Crops', *International Journal of Biotechnology* 1(1/2/3).

Levidow, L., S. Carr, R. von Schomberg and D. Wield (1996) 'Regulating Agricultural Biotechnology in Europe: Harmonisation Difficulties, Opportunities, Dilemmas', *Science and Public Policy* 23(3): 135–57.

Levidow. L., S. Carr and D. Wield (2000) 'Genetically Modified Crops in the European Union: Regulatory Conflicts as Precautionary Opportunities', *Journal of Risk Research* 3(3): 189–208.

Levidow, L., S. Carr and D. Wield (2005) 'EU Regulation of Agri-Biotechnology: Precautionary Links between Science, Expertise and Policy', *Science & Public Policy* 32(4): 261–76.

Levidow, L. and C. Marris (2001) 'Science and Governance in Europe: Lessons from the Case of Agbiotech', *Science and Public Policy* 28(5): 345–60.

Levidow, L., J. Murphy and S. Carr (2007) 'Recasting "Substantial Equivalence": Transatlantic Governance of GM Food', *Science, Technology and Human Values* 32(1).

Levidow, L. and J. Tait (1995) 'The Greening of Biotechnology: GMOs as Environment-Friendly Products' in I. Moser and V. Shiva (eds) *Biopolitics: A Feminist and Ecological Reader on Biotechnology* (London: Zed Books), pp. 121–39.

Levy, D. (1997) 'Business and International Environmental Treaties: Ozone Depletion and Climate Change', *California Management Review* 39(3).

Levy, D. and D. Egan (1998) 'Capital Contests: National and Transnational Channels of Corporate Influence on the Climate Change Negotiations', *Politics and Society* 26(3).

Levy, D. and P. Newell (2000) 'Oceans Apart? Business Responses to Global Environmental Issues in Europe and the United States', *Environment*, 42(9): 8–20.

Levy, D. and P. Newell (2002) 'Business Strategy and International Environmental Governance: A Neo-Gramscian Synthesis', *Global Environmental Politics* 2(4): 84–101.

Levy, D. and P. Newell (eds) (2005) *The Business of Global Environmental Governance* (Cambridge, MA: MIT Press).

Lijie, C. (2002) 'China', in C. Bail, R. Falkner and H. Marquard (eds) *The Cartagena Protocol on Biosafety: Reconciling Trade in Biotechnology with Environment and Development?* (London: Earthscan), pp. 160–65.

Lipskey, M. (1980) *Street-Level Bureaucracy: Dilemmas of the Individual in Public Services*, New York: Russell Sage Foundation.

Lipton, M. (1999) 'Reviving Global Poverty Reduction: What Role for Genetically Modified Plants?' Sir John Crawford Memorial Lecture, Washington: CGIAR.

Lisboa, M. V. (2002) 'Em Busca de Uma Política Externa Brasileira de Meio Ambiente: Três Exemplos e uma Exceção à Regra', *São Paulo em Perspectiva* 16(2): 44–52.

Lukes, S. (1974) *Power, a Radical View* (London: Macmillan).

Mackenzie, R. (2002) 'The International Regulation of Modern Biotechnology', *Yearbook of International Environmental Law* 13: 97–163.

Mackenzie, R. (2004) 'The Cartagena Protocol after the First Meeting of the Parties', *Review of European Community and International Environmental Law* 13(3): 270–78.

Mackenzie, R. and P. Sands (2002) 'Prospects for International Environmental Law', in C. Bail, R. Falkner and H. Marquard (eds) *The Cartagena Protocol on Biosafety: Reconciling Trade in Biotechnology with Environment and Development?* (London: Earthscan), pp. 457–66.

Majone, G. (1996) 'The European Commission as Regulator', in G. Majone (ed) *Regulating Europe* (London, UK: Routledge).

Mansour, M. and J. Bennett (2000) 'Codex Alimentarius, Biotechnology and Technical Barriers to Trade' *AgBioForum* 3(4): 1–5.

Mantegazzini, M. C. (1986) *The Environmental Risks from Biotechnology* (London: Frances Pinter).

Marquard, H. (2002) 'Scope', in C. Bail, R. Falkner and H. Marquard (eds) *The Cartagena Protocol on Biosafety: Reconciling Trade in Biotechnology with Environment and Development?* (London: Earthscan), pp. 289–98.

Mathews, J. T. (1997) 'Power Shift', *Foreign Affairs* 76(1): 50–66.

Mayr, J. (2002) 'Colombia', in C. Bail, R. Falkner and H. Marquard (eds) *The Cartagena Protocol on Biosafety: Reconciling Trade in Biotechnology with Environment and Development?* (London: Earthscan), pp. 218–29.

McAllister, L. K. (2005) 'Judging GMOs: Judicial Application of the Precautionary Principle in Brazil', *Ecology Law Quarterly* 32(1): 149–74.

McGinn, A. P. (2000) *Why Poison Ourselves? A Precautionary Approach to Synthetic Chemicals*. Worldwatch Paper 153.

Mellen, M. (2003) 'Who is Getting Fed?' *Seedling* (April). Available at http://www.grain.org/seedline/seed-03-04-3-en.cfm.

Millstone, E. and P. van Zwanenberg (2001) 'Politics of Expert Advice: Lessons from the Early History of the BSE Saga', *Science and Public Policy* 28(2): 99–112.

Millstone, E. and P. van Zwanenberg (2003) 'Food and Agricultural Biotechnology Policy: How Much Autonomy Can Developing Countries Exercise?', *Development Policy Review* 21(5–6): 655–67.

Milner, H. (1991) 'The Assumption of Anarchy in International Relations Theory', *Review of International Studies* 17(1): 67–86.

Milner, H. (1998) 'International Political Economy: Beyond Hegemonic Stability', *Foreign Policy* (Spring): 112–23.

Moeller, D. and Sligh, M. (2004) *Farmers' Guide to GMOs*. (St Paul, MN: Farmers' Legal Action Group, Inc.).

Moore, T. G. (2000) 'China and Globalization', in S. S. Kim (ed) *East Asia and Globalization* (Lanham: Rowman & Littlefield), pp. 105–31.

Moore, T. G. (2002) *China in the World Market: Chinese Industry and International Sources of Reform in the Post-Mao Era* (Cambridge, Cambridge University Press).

MSNBC (2001) *Public Concerned about Bio-Foods*. Available at http://lists.iatp.org/listarchive/archive.cfm?id=26891.

Murphy, J. and L. Levidow (2006) *Governing the Transatlantic Conflict over Agricultural Biotechnology: Contending Coalitions, Trade Liberalisation and Standard Setting* (London: Routledge).

National Academy of Sciences (1983) *Risk Assessment in the Federal Government: Managing the Process*. Committee on the Institutional Means for Assessment of Risks to Public Health, Commission of Life Sciences (Washington: National Academy Press).

National Association of Wheat Growers (NAWG) (2004) *Letter to Robert Zoellick*, portions reprinted in the NAWG Weekly Newsletter, (13 February). Available at http://www.wheatworld.org.

Nature Biotechnology (2005) 'Open Sesame' (editorial), *Nature Biotechnology* 23(6): 633.

Neumayer, E. (2005) 'Is the Allocation of Food Aid Free from Donor Interest Bias?' *Journal of Development Studies* 41(3): 394–411.

Newell, P. (2001) 'Managing Multinationals: The Governance of Investment for the Environment' *Journal of International Development* 13: 907–19.

Newell, P. (2003a) *Biotech Firms, Biotech Politics: Negotiating GMOs in India*. IDS Working Paper 201, September. (Brighton: IDS).

Newell, P. (2003b) *Domesticating Global Policy on GMOs: Comparing India and China*. IDS Working Paper 206 (Brighton: Institute of Development Studies).

Newell, P. (2003c) 'Globalization and the Governance of Biotechnology', *Global Environmental Politics* 3(2): 56–71.

Newell, P. and R. Mackenzie (2000) 'The 2000 Cartagena Protocol on Biosafety: Legal and Political Dimensions', *Global Environmental Change* 10: 313–17.

Newell, P. and R. Mackenzie (2003) 'Whose Rules Rule? Development and the Global Governance of Biotechnology', *IDS Bulletin* 35(1): 82–92.

Newell, P. and M. Paterson (1998) 'A Climate for Business: Global Warming, the State and Capital', *Review of International Political Economy* 5(4).

Nogueira, A. H. V. (2002) 'Brazil', in C. Bail, R. Falkner and H. Marquard (eds) *The Cartagena Protocol on Biosafety: Reconciling Trade in Biotechnology with Environment and Development* (London: Earthscan), pp. 129–37.

Nuffield Council on Bioethics (1999) *Genetically Modified Crops: The Ethical and Social issues*, May (London).

Nye, J. (1991) *Bound to Lead: The Changing Nature of American Power* (New York: Basic Books).

OECD (1986) *Recombinant DNA Safety Considerations* (Paris: Organization for Economic Cooperation and Development).

OECD (1992) *Safety Considerations for Biotechnology* (Paris: Organization for Economic Cooperation and Development).

Ogolla, B., M. A. Lehmann and X. Wang (2003) 'International Biodiversity and the World Trade Organization: Relationship and Potential for Mutual Supportiveness', *Environmental Policy and Law* 33(3–4): 117–32.

Ollinger, M. and J. Fernandez-Cornejo (1995) 'Innovation and Regulation in the Pesticide Industry' US Census, Economic Studies 95-14. Available at http://www.ces.census.gov/paper.php?paper=100237&PHPSESSID=f818b5c9f7ef0325949ceee16223c88.

Ollinger, M. and J. Fernandez-Cornejo (1998) 'Sunk Costs and Regulation in the U.S. Pesticide Industry' *International Journal of Industrial Organization* 16: 139–68.

Ollinger, M. and L. Pope (1995) 'Strategic Research Interests, Organizational Behavior, and the Emerging Market for the Products of Plant Biotechnology', *Technological Forecasting and Social Change* 50(1): 55–68.

Olson, M., (1965) *The Logic of Collective Action: Public Goods and the Theory of Groups*. (Cambridge, MA: Harvard University Press).

O'Neill, M. (2001) 'Modified Crops Advocate Fears Lost Chances Through Cultivation Ban', *South China Morning Post*, 18 April.

Organic Consumers Association/OCA (2001) *Countries & Regions with GE Food/Crop Bans*. Available at http://www.purefoods.org/gefood/countrieswithbans.cfm.

Oxfam America (2003) *US Export Credits: Denials and Double Standards* (Washington, D.C.: Oxfam America).

Oxfam International (2002) 'Crisis in Southern Africa', *Oxfam Briefing Paper 23*.

Paarlberg, R. L. (2000) 'The Global Food Fight', *Foreign Affairs* 79(3): 24–38.

Paarlberg, R. L. (2001) *The Politics of Precaution: Genetically Modified Crops in Developing Countries* (Baltimore: The Johns Hopkins University Press).

Paarlberg, R. L. (2003) 'Reinvigorating Genetically Modified Crops', *Issues in Science and Technology* 19(3).

Palaez, V. and W. Schmidt (2004) 'Social Struggles and the Regulation of Transgenic Crops in Brazil', in K. Jansen and S. Veldema (eds) *Agribusiness and Society: Corporate Responses to Environmentalism, Market Opportunities and Public Regulation* (London: Zed Books).

Pardo, M. (2005) *Transnacionalización del Conflicto sobre las Regalías de la Soja*. Available at www.agropecuaria.org/transgenicos/PardoSojaRegalias.htm.

Parsons, T. (1967) *Sociological Theory and Modern Society* (New York: The Free Press).

Pearce, F. (2003) 'UN is Slipping Modified Food into Aid', *New Scientist* 175(2361): 5.

Pereira, S. R. (2004) 'A Evolução do Complexo Soja e a Questão da Transgenia', *Revista de Política Agrícola* 13(2): 26–32.

Persson, T. and G. Tabellini (2000) *Political Economics: Explaining Economic Policy* (Cambridge, MA: MIT Press).

Pesticides Action Network Updates Service (PANUPS) (2001) 'Handful of Corporations Dominate Commercial Agriculture', 10 September. Available at http://www.panna.org/panna/resources/panups/panup_20010910.dv.html.

Pettauer, D. (2002) 'Interpretation of Substantial Equivalence in the EU', in *Evaluating Substantial Equivalence: A Step Towards Improving the Risk/Safety Evaluation of GMOs* (Wien: Umweltbundesamt (UBA)), pp. 15–24. Available at http://www.umwelt bundesamt.at/fileadmin/site/publikationen/CP032.pdf.

Pew Initiative on Food and Biotechnology (2002) *State Legislative Activity in 2001 Related to Agricultural Biotechnology*. Available at http://pewagbiotech.org/resources/factsheets/bills/factsheet.php3.

Pew Initiative on Food and Biotechnology (2004) 'Genetically Modified Crops in the United States' (factsheet). Available at http://pewagbiotech.org/resources/fact-sheets/display.php3?FactsheetID=2.

Pew Initiative on Food and Biotechnology (2005) *State Legislative and Local Activities Related to Agriculture Biotechnology Continue to Grow in 2003-2004*. Available at http://pewagbiotech.org/resources/factsheets/legislation/factsheet.php.

Phillips, N. (2001) 'Regionalist Governance in the New Political Economy of Development: "Relaunching" the Mercosur', *Third World Quarterly* 22(4): 565–83.

Phillips, P. W. B. and D. Corkindale (2002) 'Marketing GM Foods: The Way Forward', *AgBioForum* 5(3): 113–21.

Picciotto, S. (2002) 'Introduction: Reconceptualizing Regulation in the Era of Globalization', *Journal of Law and Society* 29: 1–11.

Pollack, A. (2001) 'Farmers Joining State Efforts Against Bioengineered Crops', *New York Times*, 24 March.

Pollack, M. A. and G. C. Shaffer (2001) 'The Challenge of Reconciling Regulatory Differences: Food Safety and GMOs in the Transatlantic Relation', in M. Pollack and G. Shaffer (eds) *Transatlantic Governance in a Global Economy* (Lanham: Rowman and Littlefield), pp. 153–78.

Pollack, M. A. and G. C. Shaffer (2005) 'Biotechnology Policy', in H. Wallace, W. Wallace and M. A. Pollack (eds) *Policy-Making in the European Union* (Oxford: Oxford University Press), pp. 329–51.

Prakash, A. (2002) 'Beyond Seattle: Globalization, the Nonmarket Environment and Corporate Strategy', *Review of International Political Economy* 9(3): 513–37.

Prakash A. and K. L. Follman (2003) 'Biopolitics in the EU and the US: Race to the Bottom or Convergence to the Top?' *International Studies Quarterly* 47(4): 617–41.

Priest, S. H., H. Bonfadelli and M. Rusanen (2003) 'The "Trust Gap" Hypothesis: Predicting Support for Biotechnology across National Cultures as a Function of Trust in Actors', *Risk Analysis* 23(4): 751–66.

Princen, S. (2004) 'Trading Up in the Transatlantic Relationship', *Journal of Public Policy* 24(1): 127–44.

Princen, T. and M. Finger (eds) (1994) *Environmental NGOs in World Politics. Linking the Global and the Local* (London: Routledge).

Purdue, D. A. (2000) *Anti-GenetiX: The Emergence of the Anti-GM Movement* (Burlington, VT: Ashgate).

Putnam, R. D. (1995 'Bowling Alone: America's Declining Social Capital', *Journal of Democracy* 6(1): 65-78

Putnam, R. D. (2002 *Democracies in Flux: The Evolution of Social Capital in Contemporary Society* (Oxford: Oxford University Press).

Ramakrishna, T. (2003) 'Development of the IPR Regime in India with Reference to Agricultural Biotechnology', Centre for Intellectual Property Rights, Research and Advocacy, National Law School of India, Bangalore. Paper for the project Globalization and the International Governance of Modern Biotechnology.

Raustila K. and D. G. Victor (2004) 'The Regime Complex for Plant Genetic Resources', *International Organization* 58(2): 277–309.

Redgwell, C. (2005) 'Biotechnology, Biodiversity and International Law', *Current Legal Problems* 58: 543–59.

Reifschneider, L. (2002) 'Global Industry Coalition', in C. Bail, R. Falkner and H. Marquard (eds) *The Cartagena Protocol on Biosafety: Reconciling Trade in Biotechnology with Environment and Development?* (London: Earthscan), pp. 273–77.

Reinalda, B. (1997) 'Private in Form and Public in Purpose: (I)NGOs as Political Actors in World Politics'. Paper for the ECPR 25th Joint Sessions of Workshops, Bern, Switzerland, March 1997.

Reuters (1999) 'Monsanto aims to plant Bt corn in China', Available at http://www.netlin..ce/gen/Zeitung/2000.

Rich, A. (1997) 'Après la vache folle, récidive sur le maïs transgénique', page 1; 'Pourquoi ce maïs transgénique et quelles garanties sanitaires', page 8; "Une décision réfléchie' ou 'Une décision dans l'urgence'"?, p. 8, *Le Soir*, 27 January, Brussels.

Ringius, L. (1997) 'Environmental NGOs and Regime Change: The Case of Ocean Dumping of Radioactive Waste', *European Journal of International Relations* 3(1): 61–104.

Risse, T., S. C. Ropp and K. Sikkink (eds) (1999) *The Power of Human Rights: International Norms and Domestic Change* (Cambridge, Cambridge University Press).

Risse-Kappen, T. (ed) (1995) *Bringing Transnational Relations Back In. Non-State Actors, Domestic Structure and International Institutions* (Cambridge: Cambridge University Press).

Roberts, M. and R. Begey (1994) 'Pesticide Producers Innovate Under Recessionary Pressure', *Chemical Week* 155(9).

Roessing, A. C. and J. J. Lazzarotto (2004) *Criação de Empregos pelo Complexo Agroindustrial da Soja* (Londrina, RS, Brazil: Embrapa Soja).

Rosenau, J. (1995) 'Governance in the Twenty-First Century', *Global Governance* 1(1).

Rosendal, G. K. (2001) 'Impacts of Overlapping International Regimes: The Case of Biodiversity', *Global Governance* 7(1): 95–118.

Rowlands, I. (1995) *The Politics of Global Atmospheric Change* (Manchester: Manchester University Press).

Rowlands, I. (2000) 'Beauty and the Beast? BP's and Exxon's Positions on Global Climate Change', *Environment and Planning C: Government and Policy* 18: 39–54.

Rugaber, C. (2002) 'U.S.-China GMO Dispute Heats Up As Senators Urge President Bush's Involvement', *International Environment Reporter* 25(14): 664–65.

Ruttan, V. (1993) 'The Politics of U.S. Food Aid Policy: A Historical Review,' in V. Ruttan (ed) *Why Food Aid?* (Baltimore: Johns Hopkins University Press).

Sabatier, P. A. and H. C. Jenkins-Smith (eds) (1993) *Policy Change and Learning: An Advocacy Coalition Approach* (Boulder, CO: Westview Press).

Safrin, S. (2002a) 'Treaties in Collision? The Biosafety Protocol and the World Trade Organization Agreements', *American Journal of International Law* 96(3): 606–28.

Safrin, S. (2002b) 'The Relationship with Other Agreements: Much Ado About a Savings Clause', in C. Bail, R. Falkner and H. Marquard (eds) *The Cartagena Protocol on Biosafety: Reconciling Trade in Biotechnology with Environment and Development?* (London: Earthscan), pp. 438–54.

Sands, P. (2003) *Principles of International Environmental Law.* 2nd ed. (Cambridge: Cambridge University Press).

Schaper, M. and S. Parada (2001) *Organismos Genéticamente Modificados: Su Impacto Socioeconómico en la Agricultura de los Países de la Comunidad Andina, Mercosur y Chile.* (Santiago de Chile: CEPAL).

Schwartz, J. (2004) 'Environmental NGOs in China: Roles and Limits', *Pacific Affairs* 77(1): 28–49.

Scott, C. (2002) 'Private Regulation of the Public Sector: A Neglected Facet of Contemporary Governance', *Journal of Law and Society* 29: 56–76.

SEMA-SP (1997) *O Mercosul e o Meio Ambiente* (São Paulo: Secretaria do Meio Ambiente).

Seshia, S. (2002) 'Plant Variety Protection and Farmers' Rights: Law-Making and the Cultivation of Varietal Control', *Economic and Political Weekly* 6 July, pp. 2741–47.

Simmons, B. A. and Z. Elkins (2004) 'The Globalization of Liberalization: Policy Diffusion in the International Political Economy', *American Political Science Review* 98(1): 171–89.

Skjaerseth, J. B. and T. Skodvin (2001) 'Climate Change and the Oil Industry: Common Problems, Different Strategies', *Global Environmental Politics* 1(4): 43–64.

Sklair, L. (2001) *The Transnational Capitalist Class* (London: Blackwell).

Skogstad, G. and E. Moore (2003) *Policy Resilience and Path Dependence: Policy Networks, Programmatic Ideas and GMO Regulatory Policies.* University of Toronto Working Paper.

Smith, E., T. Marsden and A. Flynn (2004) 'Regulating Food Risks: Rebuilding Confidence in Europe's Food?' *Environmental and Planning C: Government and Policy* 22: 543–67.

Smith, N. and C. Rugaber (2002) 'China to Issue Temporary Certificates to Allow Imports of GMOs to Continue', BioTech Watch, 8 March. Available at http://sub-script.bna.com/SAMPLES.btb.nsf.

Søgaard, V. (2001) 'Bayer AG: Chemicals and Life Sciences' *AgBioForum* 4(1): 68–73.

Stairs, D. (2000) 'Foreign Policy Consultations in a Globalizing World: The Case of Canada, the WTO and the Shenanigans in Seattle', *Policy Matters* 1(8).

Stapp, K. (2003) 'Biotech Boom Linked to Development Dollars, Say Critics', *Inter Press Service News Agency*, 3 December. Available at http://www.ipsnews.net/interna.asp?idnews=21395.

Stiles, K. (2005) 'Theories of Non-Hegemonic Cooperation', Paper presented at the annual meetings of the International Studies Association, Honolulu, Hawaii, 1–5 March, 2005.

Stoll, P.-T. (2000) 'Controlling the Risks of Genetically Modified Organisms: The Cartagena Protocol on Biosafety and the SPS Agreement', *Yearbook of International Environmental Law* 10: 82–119.

Sullivan, L. R. (ed) (1995) *China Since Tiananmen: Political, Economic, and Social Conflicts* (Armonk, NY: M.E. Sharpe).

Sullivan, L. R. (2005) 'Debating the Dam: Is China's Three Gorges Project Sustainable?' in R. S. Axelrod, D. L. Downie and N. J. Vig (eds) *The Global Environment: Institutions, Law, and Policy* (Washington, D.C.: CQ Press), pp. 244–60.

Sullivan, R. (2000) 'Biosafety Protocol Compromise', *Earth Island Journal* 15(2).

Sunstein, C. R. (2002) *Risk and Reason: Safety, Law, and the Environment* (Cambridge: Cambridge University Press).

Suttmeier, R. P. (1980) *Science, Technology and China's Drive for Modernization* (Stanford: Hoover Institution Press).

Tamiotti, L. and M. Finger (2001) 'Environmental Organizations: Changing Roles and Functions in Global Politics', *Global Environmental Politics* 1(1): 56–76.

The Ecologist (2003) 'GM Food Aid', *The Ecologist* 33(2): 46.

The United States-China Business Council (2000) Letter to Larry Combest, House of Representatives, Washington D.C., 6 January. Available at http://www.uschina.org/public/wto/usbspntr/combest.html.

Thomson, J. (2002) *Genes for Africa: Genetically Modified Crops in the Developing World* (Cape Town: University of Cape Town Press).

Thompson-Feraru, A. (1974) Transnational Political Interests and the Global Environment, *International Organization* 28(1): 31–60.

Toke, D. (2004) *The Politics of GM Food: A Comparative Study of the UK, USA and EU* (London: Routledge).

Trigo, E. J. and E. J. Cap. (2003) 'The Impact of the Introduction of Transgenic Crops in Argentinean Agriculture', *AgBioForum* 6(3): 87–94.

UDS (2002) 'China's Biotech Rules Won't Harm Trade, Veneman Says', United States Department of State, 3 July. Available at http://www.usda.gov/news/releases/2002.

UNCTAD (2004) *World Investment Report 2004: The Shift Towards Services* (Geneva: UNCTAD).

UNEP (1995) *Report of the Open-Ended Ad Hoc Group of the Experts on Biosafety*, UNEP/CBD/COP/2/7.

UNEP (1999) *Conference of the Parties to the Convention on Biological Diversity. First Extraordinary Meeting of the Open-Ended Ad Hoc Working Group on Biosafety*, UNEP/CBD/ ExCOP/1/2, 15 February 1999.

UNEP (2003) *Reports of the Intergovernmental Committee for the Cartagena Protocol on Biosafety*, UNEP/CBD/BS/COP-MOP/1/3/Add.1/24 May 2003 (first meeting), UNEP UNEP/CBD/BS/COP-MOP/1/3/Add.2/24 May 2003 (second meeting), UNEP/CBD/BS/COP-MOP/1/3/Add.3/24 May 2003 (third meeting))

UNEP (2004) *Report of the First Meeting of the Conference of Parties to the Convention on Biological Diversity Serving as the Meeting of the Parties to the Cartagena Protocol on Biosafety*, UNEP/CBD/BS/COP-MOP/1/15/14, April 2004.

UNEP (2005) *Report of the Second Meeting of the Conference of Parties to the Convention on Biological Diversity Serving as the Meeting of the Parties to the Cartagena Protocol on Biosafety*, UNEP/CBD/BS/COP-MOP2/15, 6 June 2005.

UNEP Chemicals website http://www.chem.unep.ch/pops/newlayout/infpopsalt.htm.

United States Agency for International Development (USAID) (2003a) 'CABIO: Mobilizing Science and Technology to Reduce Poverty and Hunger', Press Release 2003-063, Washington, D.C. Available at http://www.usaid.gov.

United States Agency for International Development (USAID) (2003b) 'Educating Scientists for Management', Press Release 2003-070, Washington, D.C. Available at http://www.usaid.gov.

USAID (1995) *Food Aid and Food Security Policy Paper* (PN-ABU-219) (Washington, D.C.: USAID).

USAID (2003c) 'United States and Food Assistance', *Africa Humanitarian Crisis*, 3 July. Available at www.usaid.gov/about/africafoodcrisis/bio_answers.html#8.

USDA (2001) *U.S. Agricultural Trade*. Available at http://www.ers.usda.gov/BRIEF-ING/AgTrade/usagriculturaltrade.htm.

USDA (2003) *The National Organic Program*. Available at http://www.ams.usda.gov/nop/Q&A.html.

USDA (2005) *Agricultural Biotechnology: Adoption of Biotechnology and Its Production Impacts*. Available at http://www.ers.usda.gov/Briefing/biotechnology/ chapter1.htm.

USEPA (2001) 'Biotechnology Corn Approved for Continued Use', Press Release, 16 October. Available at www.usepa.gov.

Uvin, P. (1992) 'Regime, Surplus and Self-Interest: The International Politics of Food Aid,' *International Studies Quarterly* 36(3): 293–312.

Victor, D. G. and C. F. Runge (2002) 'Farming the Genetic Frontier', *Foreign Affairs* 81(May–June): 107–21

Vig, N. J. and M. G. Faure (eds) (2004). *Green Giants? Environmental Policies of the United States and the European Union* (Cambridge, MA: MIT Press).

Vogel, D. (1995) *Trading Up. Consumer and Environmental Regulation in a Global Economy* (Cambridge, MA: Harvard University Press).

Vogel, D. (1997) 'Trading Up and Governing Across: Transnational Governance and Environmental Protection', *Journal of European Public Policy* 4(4): 556–71.

Vogel, D. (2003) 'The Hare and the Tortoise Revisited: The New Politics of Consumer and Environmental Regulation in Europe', *British Journal of Political Science* 33(4): 557–80

Vogler, J. (2003) 'Taking Institutions Seriously: How Regime Analysis Can Be Relevant to Multilevel Environmental Governance', *Global Environmental Politics* 3(2): 25–39.

Vogler, J. (2005) 'The European Contribution to Global Environmental Governance', *International Affairs* 81(4): 835–50.

Vogt, D. U. and M. Parish (1999) *Food Biotechnology in the United States: Science, Regulation, and Issues*. Available at http://www.cnie.org/nle/st41.html.

Vogt, D. U. and M. Parish (2001) *Food Biotechnology in the United States: Science, Regulation, and Issues*, RL 30198. Available at http://cnie.org/NLE/CRSreports/science/st-41.pdf.

Wallerstein, M. (1980) *Food for War—Food for Peace* (Cambridge, MA: MIT Press).

Walsh, D. (2000) 'America Finds Ready Market for GM Food—The Hungry', *The Independent* (London), 30 March: 18.

Wang, X. (2004) 'Challenges and Dilemmas in Developing China's National Biosafety Framework', *Journal of World Trade* 38(5): 899–913.

Wang, X. and G. Wiser (2002) 'The Implementation and Compliance Regimes under the Climate Change Convention and Its Kyoto Protocol', *Review of European Community and International Environmental Law* 11(2): 181–98.

Washington Post (1999) 'Biotech: Yes or No?' *Washington Post,* 16 October, p. A 19.

Waste, R. J. (ed) (1986) *Community Power. Directions for Future Research* (Beverly Hills: SAGE).

Werner, K. (2001a) 'EPA Conditionally Reapproves Biotech Corn for Seven Years; Additional Data Requested', *BioTech Watch*, 17 October.

Werner, K. (2001b) 'Industry Says More Regulatory Scrutiny of Biotech Crops Could Spur Chemical Use', *BioTech Watch*, 2 October.

Werner, K. (2001c) 'Industry Supports EPA Biotech Exemption, Citing Safe Crop History, Regulatory Burden', *BioTech Watch*, 11 October.

West, D. and B. A. Loomis (1999) *The Sound of Money: How Political Interests Get What They Want* (New York: Norton).

Westerheijden, D. (1987) 'The Substance of a Shadow. A Critique of Power Measurement Methods', *Acta Politica* (1): 39–59.

WFP (2002). *Policy on Donations of Foods Derived from Biotechnology (GM/Biotech foods)*, WFP/EB, 3/2002/4-C, 14 October.

WFP (2003) *Policy on Donations of Foods Derived from Biotechnology*, WFP/EB.A/2003/5-0B/Rev.1.

White House (2000) *Clinton Administration Agencies Announce Food and Agricultural Biotechnology Initiative*, 3 May. Available at http://vm.cfsan.fda.gov/~lrd/whbio53.html.

Willets, P. (ed) (1982) *Pressure Groups in the Global System* (London: Frances Pinter).

Willets, P. (1996) 'From Stockholm to Rio and Beyond: The Impact of the Environmental Movement on the United Nations Consultative Arrangements for NGOs', *Review of International Studies* 22(1): 57–80.

Winston, M. L. (2002) *Travels in the Genetically Modified Zone* (Cambridge, MA: Harvard University Press).

World Intellectual Property Report (2005) 'Monsanto Mulls WIPO Action Against Argentina over Soybeans', *World Intellectual Property Report* 19(3).

Wright, B. G. (2000) 'Environmental NGOs and the Dolphin-Tuna Case', *Environmental Politics*, 9(4): 82–103.

WTO (2001) *Matrix of Multilateral Environmental Agreements*. WTO Doc WT/CTE/W/160.Rev.1, Committee on Trade and the Environment, 14 July (Geneva WTO).

WTO (2004) *World Trade Report 2004: Exploring the Linkage between the Domestic Policy Environment and International Trade* (Geneva: World Trade Organization).

WTO AB (1998) World Trade Organisation Appellate Body, 'EC – Hormones Appellate Body Report: EC Measures Concerning Meat and Meat Products (Hormones), WT/DS26/AB/R, WT/DS48/AB/R', adopted 13 February, DSR 1998: I, 135.

WWF (2003) *Sustainability Assessment of Export-Led Growth in Soy Production in Brazil* (Brasilia: WWF-Brazil).

Wynne, B. (2001) 'Creating Public Alienation: Expert Cultures of Risk and Ethics on GMOs' *Science as Culture* 10(4): 445–81.

Xinhua (2002) 'Vice-Premier Hails Sustainable Development' *Xinhua News Agency*, 25 November.

Yamin, F (2002) *IPRs, Biotechnology and Food Security*, Paper for FIELD/IDS Project on the International Governance of Modern Biotechnology.

Yang, G. (2005) 'Environmental NGOs and Institutional Dynamics in China', *The China Quarterly* (181): 46–66.

Yang, W. (2003) 'Regulation of Genetically Modified Organisms in China', *Review of European Community and International Environmental Law* 12(1): 99–108.

Young, A. (2003) 'Political Transfer and 'Trading Up'? Transatlantic Trade in Genetically Modified Foods and U.S. Politics', *World Politics* 55(July): 457–84.

Yu III, V. P. B. (2001) 'Compatibility of GMO Import Regulations with WTO Rules', in E. B. Weiss and J. H. Jackson (eds) *Reconciling Environment and Trade* (Ardsley, NY: Transaction Publishers).

Zedan, H. (2002) 'The Road to the Biosafety Protocol', in C. Bail, R. Falkner and H. Marquard (eds) *The Cartagena Protocol on Biosafety: Reconciling Trade in Biotechnology with Environment and Development?* (London: Earthscan), pp. 23–33.

Zhao, J. and L. Ortolano (2003) 'The Chinese Government's Role in Implementing Multilateral Environmental Agreements: The Case of the Montreal Protocol', *The China Quarterly* (167): 708–25.

Zürn, M. and J. T. Checkel (2005) 'Getting Socialized to Build Bridges: Constructivism and Rationalism, Europe and the Nation-State', *International Organization* 59(4): 1045–79.

Zweig, D. (2002) *Internationalizing China: Domestic Interests and Global Linkages* (Ithaca: Cornell University Press).

Index